DOING BUSINESS IN A NEW CLIMATE
A GUIDE TO MEASURING, REDUCING AND OFFSETTING GREENHOUSE GAS EMISSIONS

Paul Lingl, Deborah Carlson
and the David Suzuki Foundation

publishing for a sustainable future

LONDON • WASHINGTON, DC

First published by Earthscan in the UK and USA in 2010

HB ISBN: 978-1-84407-907-0
PB ISBN: 978-1-84407- 908-7

Design and production: Arifin Graham, Alaris Design
Illustrations: IAASTD/Ketill Berger, UNEP/GRID-Arendal (p. x); Marcia Ritz (p. 10); Soren Henrich (pp. 15, 29, 47–49); Roger Handling (p. 42); others as credited.
Photographs: Kent Kallberg (pp. iii, iv, 20); courtesy TransAlta Wind (p. iii); Interface, Inc. (p. 5); Hudson's Bay Company (p. 32); Resort Municipality of Whistler (p. 39); Paul Lingl (p. 44); Franz Patzig (p. 53); all others iStockphoto.com

Unless otherwise specified, all weights and measures are given in metric units, and all currency figures are in US dollars.

This guide offers general guidance for businesses interested in greenhouse gas (GHG) management. Businesses that undertake GHG management programmes may wish to acquire specific training in GHG management, or work with accredited GHG consultants. Mention of specific companies, products or services in this guide does not necessarily imply endorsement of these companies, products or services by the authors, the David Suzuki Foundation or Earthscan.

This book was made possible by the generous support of Vancity, the Stephen R. Bronfman Foundation and the Bullitt Foundation.

Royalties from the sale of this book will go to support the work of the David Suzuki Foundation.

David Suzuki Foundation
2211 West 4th Avenue, Suite 219
Vancouver, BC, Canada V6K 4S2
www.davidsuzuki.org
Tel +604 732 4228
Fax +604 732 0752

Earthscan Ltd, Dunstan House, 14a St Cross Street, London EC1N 8XA, UK
Earthscan LLC, 1616 P Street, NW, Washington, DC 20036, USA
Earthscan publishes in association with the International Institute for Environment and Development

For more information on Earthscan publications, see www.earthscan.co.uk or write to earthinfo@earthscan.co.uk

A catalogue record for this book is available from the British Library

Library of Congress Cataloging-in-Publication Data has been applied for.

At Earthscan we strive to minimize our environmental impacts and carbon footprint through reducing waste, recycling and offsetting our CO_2 emissions, including those created through publication of this book. For more details of our environmental policy, see www.earthscan.co.uk.

This book was printed in Malta by Gutenberg Press.
The paper used is FSC certified and the inks are vegetable based.

You are invited to provide feedback on this book, and share your successes and challenges with GHG management, by emailing: business@davidsuzuki.org

The message from scientists is clear: we must take action now to reduce the greenhouse gas emissions we produce if we are to avoid runaway climate change.

The good news is that the challenge of reducing our emissions also offers an opportunity for innovation and prosperity that will not only protect the environment, but also strengthen our economy.

All around the world I encounter widespread concern about climate change, and I am encouraged by the many business leaders who tell me that they want to find ways to reduce their climate impact. These businesses are looking to the future by demonstrating that solutions to climate change are possible, and even profitable.

It is my hope that this guide will help businesses and other organizations start along the path to reducing their emissions, and realize the many benefits of action.

Dr. David Suzuki
CO-FOUNDER, DAVID SUZUKI FOUNDATION

Businesses from many different sectors play a role in driving the economy. One thing they all have in common is that they produce greenhouse gas emissions, including from electricity use, transportation of goods, fuels used in manufacturing, employee travel and a variety of other sources.

Fortunately, there are opportunities for businesses of all sizes and types to reduce their emissions, and businesses are discovering that going green isn't just good for the planet, it's good for the bottom line. Cutting waste – such as unnecessary energy use – means saving money. As well, the experience of many companies has shown that taking action on climate change can lead to a variety of additional benefits, from brand enhancement to loyal and motivated employees.

At the David Suzuki Foundation we are frequently contacted by businesses that want to take action, but aren't sure where to begin. That's why we created this guide. It explains how to measure, reduce and offset emissions, and also provides guidance on developing a communications strategy around these initiatives.

The guide also includes real-life examples of businesses from around the world that are already taking a leadership role when it comes to reducing their emissions, and implementing innovative greenhouse gas management programmes. Their experience and success can be used as a blueprint – and inspiration – for other businesses to follow.

Peter Robinson
CEO, DAVID SUZUKI FOUNDATION

DOING BUSINESS IN A NEW CLIMATE

TABLE OF CONTENTS

GETTING STARTED

MEASURING GHGs

REDUCING GHGs

OFFSETTING GHGs

COMMUNICATING

MOVING FORWARD

Acknowledgements

The authors would like to acknowledge the helpful input of the following individuals: Tom Baumann, Ted Battison, Brad Chapman, Michael Contardi, Josephine Coombe, Wesley Gee, Arifin Graham, Chris Higgins, L. E. Johannson, Myrna Khan, Greg Kiessling, Jacqueline Kuehnel, Julian Lee, Sheldon Leong, Darryl Luscombe, Matt McCulloch, Hillary Marshall, Sarah Marchildon, Erin Meezan, Wren Montgomery, David Moran, Sean Pander, Amanda Pitre-Hayes, Samantha Putt del Pino, Dorit Shackleton, Ginny Stratton, Magdalena Szpala, Hamish van der Ven, Bob Willard, Graham Willard, and Brian Yourish.

We also thank, from the David Suzuki Foundation: Morag Carter, Nicholas Heap, Ryan Kadowaki, Ian Bruce, Randi Kruse, Gail Mainster, Dale Marshall, Kristen Ostling, Peter Robinson, Elois Yaxley and Dr. David Suzuki for their input and assistance with the guide.

How to use this guide

This guide describes best practices in greenhouse gas management, and uses many real-life examples to illustrate the options available to businesses to reduce their climate impact and improve their bottom line. The numbered images at right show the key elements of a greenhouse gas management programme, each of which corresponds to one part of this guide. You can navigate between Parts 1 to 6 using the coloured tabs found on all right-hand pages.

To begin managing GHG emissions, you are encouraged to read Part 1 of the guide, *Getting Started*, which will help you develop a business case for action and mobilize resources. The next steps are flexible. Some businesses might begin by measuring their emissions, while others might go directly to reductions, for example. Regardless of your focus, Part 5, *Communicating Effectively* will provide useful information on how to communicate your GHG management programme to stakeholders.

Each part concludes with helpful resources. Additional resources, a glossary, and index are found at the back of the guide.

1 GETTING STARTED

2 MEASURING GHGs

3 REDUCING GHGs

4 OFFSETTING GHGs

5 COMMUNICATING

6 MOVING FORWARD

Introduction

All businesses, large and small, from the industrial sector to the service industry, produce greenhouse gas emissions. Heating and cooling office space, powering electronic equipment, transporting goods, business travel and manufacturing processes are just a few of the many activities that produce greenhouse gas emissions and contribute to climate change.

Around the world, a growing number of businesses are already taking steps to manage their greenhouse gas emissions and reduce their climate impact, often as part of broader initiatives to green their operations. These businesses are finding that there is a strong business case for managing their greenhouse gas emissions, including cost savings, brand enhancement and other forms of competitive advantage.

Because no two businesses face the same risks or opportunities with respect to climate change, there are many options when it comes to greenhouse gas (GHG) management. Some businesses have chosen to measure their emissions, and then make reductions. Others are also choosing to make a product or service, or even their entire operations, carbon neutral. A common theme is that these leading businesses are becoming more efficient, taking advantage of new markets, and helping to define how businesses will be successful in a carbon-constrained world.

This guide is designed to help businesses take action to reduce their climate impact, and at the same time reap some of the many related benefits. Without attempting to prescribe a single formula, it discusses how to define a business case and set goals for managing emissions based on the risks and opportunitiesfacing each business. The guide then identifies the key elements of a GHG management programme, including measuring and reducing GHG emissions, and explores the best practices for each. Next is an introduction to carbon off-

Managing greenhouse gases

While the concept of greenhouse gas (GHG) emissions might be new to some businesses, it's important to remember that most GHG emissions will likely result from fuel and energy consumption, and GHG management will usually mean measuring and and reducing emissions from these sources.

sets and carbon neutral initiatives, as many businesses are seeking guidance on this relatively new subject. This is followed by advice on developing an effective communications strategy as part of a greenhouse gas management programme.

The guide concludes with some ideas about how to overcome typical challenges that businesses may face. Throughout, case studies and examples from businesses and other organizations highlight achievements and illustrate innovative solutions. Many businesses that decide to manage their greenhouse gas emissions will be working with minimal expert assistance, and this guide not only outlines best practices, but also provides a number of tools, a glossary of terms, and links to further resources.

While the primary target audience of this guide is the business community, many of the greenhouse gas management practices explored also apply to other organizations that wish to reduce their climate impact, including government agencies, municipalities, non-governmental organizations, educational institutions, event organizers and others.

Climate change impacts on the economy

Climate change, also known as global warming,[1] is one of the most serious challenges the world faces. Greenhouse gases from human activities like electricity generation, manufacturing, transportation and land-use change are accumulating in the atmosphere, where they act like a heat-trapping blanket that is warming the Earth's climate.

Global warming is already creating physical impacts that have serious economic consequences. For example, warming oceans are threatening some fisheries, and warmer winters have already led to insect infestations in northern forests, resulting in billions of dollars in economic losses. Future risks include sea-level rise and shoreline erosion in coastal areas, and severe storms that will damage property and infrastructure. In some areas droughts and wildfires are expected. Regions dependent on spring run-off for their water supply will have water shortages at the same time that higher summer temperatures increase demand, jeopardizing agriculture, tourism and other industries. Further economic costs are likely to result from uncertainty in the financial sector, and lost productivity due to climate change-related disruptions.[2]

DIAGRAM 1:
GLOBAL GHG EMISSION
SOURCES, 2004

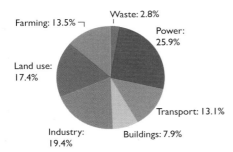

Waste: 2.8%
Farming: 13.5%
Power: 25.9%
Land use: 17.4%
Transport: 13.1%
Industry: 19.4%
Buildings: 7.9%

SOURCE
UNEP/GRID-Arendal Maps and Graphics Library. http://maps.grida.no/go/graphic/greenhouse-gas-ghg-emissions-by-source-2004, IPCC, Working Group 1, 2007.

The good news is that we have the means, if we act now, to take action and reduce our greenhouse gas emissions, and lessen the future impacts of climate change. According to Nicholas Stern, author of the most comprehensive assessment of the economics of climate change ever undertaken, we only need to invest about 2% of GDP per year in reducing our greenhouse gas emissions to avoid a reduction of up to 20% in GDP due to the impacts of climate change.[3] To put this in context, Lord Stern has observed that if climate change is not addressed we can expect economic consequences far more severe than the most recent economic downturn, which was itself the result of ignoring risks in global financial systems.[4]

Around the world, governments are beginning to respond to the global warming challenge by putting in place policies that aim to shift our economies away from a reliance on burning fossil fuels and emitting greenhouse gases. Most businesses will likely be affected by these policies in some way.

However, growing numbers of businesses are already getting started, and demonstrating that climate change is not only a risk that needs to be addressed, but also an opportunity for innovation and increased profitability.

Putting a price on carbon

Climate change has been called the 'world's greatest market failure', because releasing GHG emissions into the atmosphere has been free, and there has been little economic incentive for alternatives. A carbon price could help level the playing field between renewable energy sources, for example, and fossil fuels like oil and coal.

Governments can price carbon directly, in the form of carbon taxes, or indirectly, by setting limits on emissions with penalties for exceeding them. Another approach is to use the market to determine the price, through emissions trading: emitters are assigned permits for set amounts of GHG emissions, but if they reduce emissions below the amounts allowed they can sell their extra permits to other emitters.

SOURCE
www.occ.gov.uk/activities/stern.htm

Major greenhouse gases caused by human activity and subject to international regulation include:

1. Carbon dioxide (CO_2) is the main contributor to climate change, especially through the burning of fossil fuels like coal, oil and gas, and also as a result of deforestation and other land-use changes.

2. Methane (CH_4) is produced naturally when vegetation is burned or rotted in the absence of oxygen, but large additional amounts of methane are released by cattle farming, landfills, rice farming and the production of oil and gas.

3. Nitrous oxide (N_2O) is released by chemical fertilizers and burning fossil fuels.

4. Hydrofluorocarbons (HFCs) are chemical by-products, and are also used in some types of refrigeration equipment.

5. Perfluorocarbons (PFCs) are manufactured chemical compounds used for a variety of medical and other applications.

6. Sulphur hexafluoride (SF_6) is a manufactured compound used in some specialized applications, like insulation for high-voltage electrical equipment.

SOURCE
www.ipcc.ch

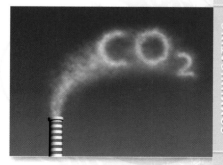

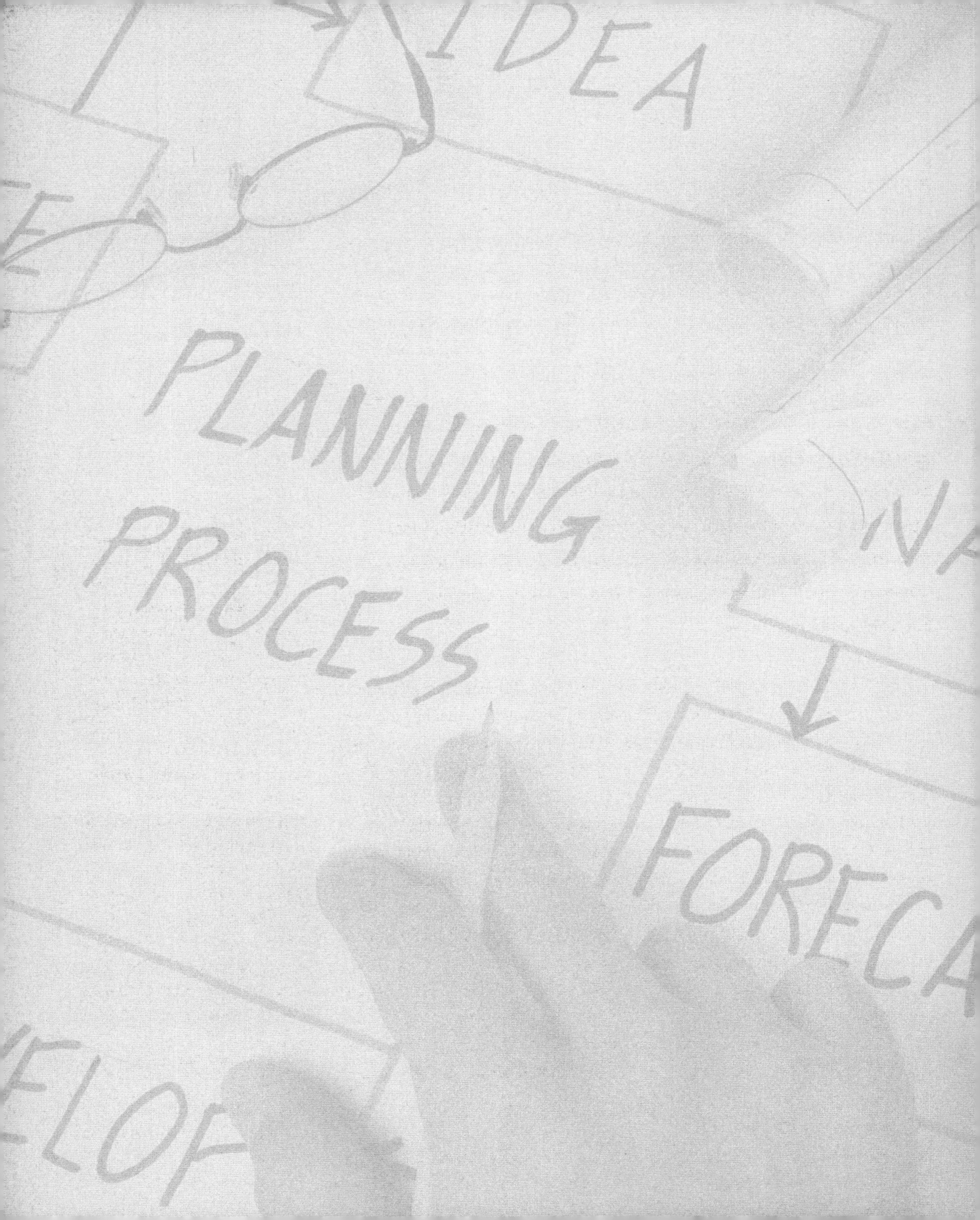

IDEA

PLANNING
PROCESS

FORECA

ELOP

GETTING STARTED

MEASURING GHGs

REDUCING GHGs

OFFSETTING GHGs

COMMUNICATING

MOVING FORWARD

PART I

Getting Started: Planning for the Success of the GHG Management Programme

Part I looks at how to design and implement a greenhouse gas (GHG) management programme for businesses. It begins by looking at how to define a business case for managing GHG emissions by examining common risks and opportunities faced by businesses. Next steps include establishing goals, obtaining buy-in and commitment, creating a climate leadership team and allocating resources.

A ll businesses and organizations, regardless of their size or sector, can manage and reduce their impact on the climate. Generally, this can be approached like any business strategy – by assessing the risks and opportunities, setting goals, assigning staff and resources, developing and implementing a plan to achieve those goals, and monitoring the results.

For businesses that already have broad sustainability programmes, managing GHG emissions can fit well into existing efforts, or, alternatively, it can be the first step towards reducing their overall environmental impact. The best-practice model of greenhouse gas management outlined in this guide involves a systematic approach, but there is still considerable flexibility for businesses to tailor a GHG management programme to their own needs.

Some businesses will choose to start with a full-scale GHG management programme, and begin by measuring emissions from their entire operations, whereas others may initially focus on the development of a single carbon neutral product. Still other businesses will set a reduction target for a particular department or business unit, or for a single category of emissions, like electricity use, or transportation.

Green initiative generates strong returns

In 2005, **GE**, one of the world's largest energy technology companies, launched a new initiative – Ecomagination – focusing on the development of clean energy solutions such as wind and solar energy, hybrid locomotives, lighter and stronger materials and efficient lighting.

By the end of 2006 Ecomagination had already produced $12 billion in revenue, with an additional $50 billion expected from pending orders and commitments. GE 'has never had an initiative that has generated better financial returns so quickly', according to Vice President Lorraine Bolsinger. Ecomagination also includes a commitment by GE to reduce its overall GHG emissions by 2012.

SOURCES
www.america.gov/st/energy_English/2007/
Septem-ber/20070914164754esnamfu
ak0.530987.html
www.ecomagination.com

Because there are many options when it comes to GHG management, it is important to do some initial planning to decide how best to proceed. Below are four steps that most businesses will find useful when creating a GHG management programme: (1) define the business case; (2) establish goals; (3) obtain buy-in and active commitment to the goals from leaders in the organization; and (4) create a climate leadership team and allocate funds to the programme.

1. Define the business case

Businesses undertake initiatives for two basic reasons: to mitigate risks or to capture opportunities – or both. Therefore, most businesses will find it helpful to begin with an analysis of the climate change issues that are likely to affect them, and their associated risks and opportunities, and then use this analysis to define their organization's business case, or rationale, for implementing a GHG management programme. Even businesses that have an initial idea of the goals for their GHG management programme will benefit from such an analysis.

At the outset, a business may not have enough information for a full analysis, but a preliminary assessment will assist in developing an initial goal or goals for the GHG management programme. As it proceeds and has better information, a business will be able to perform a more informed analysis of the risks and opportunities it faces.

As a starting point, some of the most common risks and opportunities related to climate change are summarized in Table 1 on the next page. Each business should take into consideration its own individual circumstances and then review these risks and opportunities. In doing so it may be helpful to ask the following questions:

1. **Will the company need to respond to climate change-related risks with respect to any of the issues below?**
2. **How long will it take the company to respond if necessary?**
3. **Is there a competitive advantage or other opportunity in being proactive?**

The Carbon Disclosure Project

The growing importance of climate change issues for businesses and investors is illustrated by the rapid growth of the **Carbon Disclosure Project** (CDP). This not-for-profit organization currently represents institutional investors that have a combined $55 trillion in assets under their management. On their behalf, CDP sends questionnaires to the world's largest corporations (3,700 in 2009) and asks for information about their greenhouse gas emissions data and the business risks and opportunities they face in relation to climate change. The CDP makes the responses to these questionnaires available online.

SOURCE
www.cdproject.net

Opportunities in green technology

According to leading venture capitalist, **John Doerr**, developing clean energy sources is 'the largest economic opportunity of the 21st century'. His firm has already invested hundreds of millions of dollars into green tech start-ups, such as innovative power plants, affordable solar cells for roofs and high-performance plug-in hybrid cars.

SOURCE
Testimony before US Senate Committee on Environment and Public Works, 7 January 2009

TABLE I: RISKS AND OPPORTUNITIES AS DRIVERS FOR BUSINESS ACTION ON CLIMATE CHANGE

RISKS FROM INACTION ON CLIMATE CHANGE	ISSUES FOR BUSINESSES	OPPORTUNITIES FROM ACTION ON CLIMATE CHANGE
• Continued exposure to high fuel and energy costs	**FUEL AND ENERGY COSTS**	• Cost savings from reduced fuel and energy consumption as a result of reducing GHG emissions • Improved operational efficiencies, e.g. through better fleet management
• Being a target of public campaigns singling out businesses that do not take action to reduce their climate impact	**REPUTATION**	• Brand enhancement – showing leadership on climate change can increase visibility in the marketplace and attract new customers
• Increased challenges recruiting new employees for businesses with poor records on climate change action • Higher employee costs related to lower productivity and more employee turnover as a result of employee dissatisfaction with company failure to take action	**EMPLOYEES**	• Attracting new employees looking for companies with strong sustainability programmes • Motivating employees, building loyalty and promoting employee innovation with climate change action • Enhancing employee wellness and increasing productivity through measures that also save energy (e.g. use of natural lighting, HVAC upgrades)
• Investor concern about climate change risk exposure and company inaction • Shareholder resolutions demanding specific measures to address climate change	**INVESTORS**	• Attracting new investors who want to invest in progressive, well-managed companies • Meeting corporate social responsibility goals
• Carbon taxes and other measures leading to increased fuel and energy costs • Requirements to meet energy efficiency standards for buildings and vehicles • Limits on emissions for large GHG emitters	**REGULATIONS**	• Benefitting from government incentive programmes for voluntarily reducing GHG emissions • Flexibility to choose a course of action – likely more cost-effective than waiting to be regulated • Early movers may be able to influence the shape of future regulations
• Losing customers who switch away from goods, services and technologies that are GHG-intensive	**PRODUCTS, SERVICES AND TECHNOLOGIES**	• Taking advantage of the growing demand for climate-friendly products and services
• Exposure to higher shipping costs due to higher fuel costs • Costs of GHG-intensive production by suppliers being passed along to the company	**SUPPLY CHAIN**	• Choosing suppliers with low-carbon products and services can reduce a company's upstream GHG emissions and save money at the same time • Managing transportation in the supply chain can reduce fuel consumption and GHGs, and lower costs

GETTING STARTED

MEASURING GHGs

REDUCING GHGs

OFFSETTING GHGs

COMMUNICATING

MOVING FORWARD

In addition to the risks and opportunities listed in Table 1, which are related directly to a business's own production of GHG emissions, some businesses will face direct or indirect physical risks due to climate change – for example, property holders might be affected by rising sea levels, and insurers by increased losses due to extreme weather events. Mitigating these physical risks is not addressed in this guide, but exposure to these types of risks can serve as further justification for implementing a greenhouse gas management programme.

After considering which risks and opportunities are applicable to its particular situation, a business can begin to define its case for taking action. For example, 'Business X faces exposure to carbon regulations and the rising cost of fuel, so it needs to take action to reduce its GHG emissions.'

2. Establish goals

Once a business has defined its own business case for managing GHG emissions, the next step is to formulate specific goals for its GHG management programme to help make the business case a reality.

These should be overarching goals, perhaps even visions, of where the company would like to be within a certain time frame. These goals will guide the development of plans to measure emissions and reduce the climate impact of the organization. Examples of possible goals for businesses include:

- A delivery business may decide to reduce its transportation emissions by 50%.
- A retailer may decide to make its entire operations carbon neutral.
- A bank may plan to reduce its electricity use by 35%.
- A manufacturer may decide to obtain half its energy from renewable sources, and to develop innovative low-carbon products for new markets.

Of course, more than one goal is possible, but whatever goal or goals are chosen, it is important that they are clear, and that they are taken seriously. In most organizations, an emphasis on short-term cost-effectiveness and market share tend to drive decision-making, and the goals of the GHG management programme can easily be forgotten. Having a strong business case for managing GHG emissions will therefore help ensure the goals of the GHG management are achieved.

To reinforce its goals, a business can create a formal statement that outlines its commitments. Over time, the goals can be revised and re-articulated as new information is collected, expertise developed, and risks and opportunities are reassessed.

GETTING STARTED

MEASURING GHGs

REDUCING GHGs

OFFSETTING GHGs

COMMUNICATING

MOVING FORWARD

CASE STUDY

Interface, Inc.
THE IMPORTANCE OF EXECUTIVE COMMITMENT AND LEADERSHIP

Interface, Inc. is the world's largest manufacturer of modular carpet, with more than 3,500 employees, and revenues of nearly $1 billion. It is also a leader in environmentally responsible manufacturing.

Ray Anderson (pictured) founded Interface in 1973, but it wasn't until the 1990s that he turned his focus to the environment. Customers had begun asking him about Interface's environmental impact, and in 1994 he read *The Ecology of Commerce* by Paul Hawken.

'It explained how the Earth's living systems were in decline and how industries like mine were the biggest culprits,' said Mr. Anderson. 'I was inspired to make a change.' As a result, he decided to make environmental sustainability one of his company's goals.

Interface has come a long way since 1994, and owes much of its success to Anderson's bold sustainability goals and leadership. He has led by example, swapping his Bentley for a Prius, but also through dialogue, travelling company-wide to encourage all associates to take part in the transformation of the company. He created a Global Sustainability Council to bring together managers from Interface operations around the globe and foster internal dialogue around sustainability.

The passion of Anderson's vision has inspired employees, and many Interface innovations can be traced back to a motivated company culture that encourages staff to seek out leading-edge solutions.

For example, Interface was the first company in its sector to voluntarily measure and report its annual GHG emissions publicly, and was a founding member of the EPA Climate Leaders programme. It was also the first company to offer a third-party verified climate neutral product to its customers – Cool Carpet™ – and all of Interface's products in North America are now climate neutral.

Interface also seeks to empower its employees to take action on climate change in their personal lives. Programmes like Cool Co2mmute™ and Cool Fuel™ enable Interface employees to calculate and easily offset their own carbon footprints.

Innovative technologies have also helped the company dramatically shrink its energy consumption and GHG emissions. Meticulous monitoring of energy use allows the company to strategically target improvement opportunities, ranging from replacing electrical lighting with natural light to installing highly efficient motors in its plants. Interface further reduces its climate impact by using renewable energy from solar arrays and landfill gas. By the end of 2008, Interface had successfully decreased its net absolute GHG emissions by 71% from its baseline year of 1996.

Most recently, Interface's ongoing commitment to sustainability has resulted in Mission Zero™, the company's promise to eliminate its negative impacts by 2020.

FOR MORE INFORMATION
www.interfacesustainability.com

Creating a green team
The **IKEA Group's** Social and
Environmental Affairs team is
comprised of nine people and
its manager, Thomas Bergmark,
reports directly to the president and
CEO. 'At a very early stage, the
board and the CEO recognized
how important these issues are to
a business like IKEA,' Bergmark says.
'It's a great value having the role
report to the CEO because there is
no question in the organization as
to whether this is important or not.
It's a very clear signal from the board
and from the CEO.'

SOURCE
www.edelman.com/expertise/
practices/csr/documents/
EdelmanCSR020508Final_000.pdf

3. Obtain buy-in and active commitment to the goals from leaders in the organization

An absolutely critical success factor for any GHG management programme is buy-in from key players like business owners, executives, board members, and senior managers. Their commitment and involvement will drive the message that the programme is a priority both within the company and externally. In particular, these leaders should specifically endorse and promote the goals developed in Step 2. Their commitment will need to be reiterated and reinforced over time, to help ensure continuing momentum for the programme.

Other important players include managers and the employees across the organization who will eventually be involved in implementing the GHG management programme. Less senior employees can also be pivotal to the success of the programme – like the custodian who is responsible for turning off all the lights at night. Both management and employee support can be increased if recognition and management systems are adjusted to take into account the goals of the programme.

4. Create a climate leadership team and allocate funds

Keeping in mind the goals of the GHG management programme discussed above, it is necessary to determine who will lead the programme. Some things to consider when designating the climate leadership team:

- Ensure that the team has adequate authority and representation from all key departments to ensure both buy-in and effective implementation of the programme.
- Other staff members need to be aware of the team and its activities, and its support from senior management. This matters even at early stages, because the team will likely be collecting information and data about GHG emissions from different parts of the organization. Good communication skills will also be important.
- The team will need basic knowledge about GHG management. Training, possibly through workshops or online courses, or through a consultant, may be useful.

When should a business hire a consultant to assist with its GHG management programme?

1. LACK OF EXPERTISE

A consultant can carry out the first emissions inventory, simply check results and/or train the employees of the organization to do the inventory in the future. Consultants can also perform specialized activities, like carrying out an energy audit, and suggest reduction initiatives, along with information about payback, financing and government and utility incentive programmes.

2. LACK OF TIME

If staff are too busy, a consultant can help fill in the gaps, or perform most tasks associated with the programme.

3. THE NEED FOR VERIFICATION

If independently verified results are important (e.g. if the business is making a public announcement about its initiative), consultants can be hired to plan and carry out the emissions inventory, or alternatively, and more affordably, just to verify the results.

4. POTENTIAL COSTS OF MISTAKES

The credibility of data becomes important when money is at stake – for example, in the form of penalties, capital investment to comply with regulations, or costs to purchase carbon offsets.

- The size and composition of the team will likely change over time as the GHG management programme develops and the organization gains a better understanding of the activities needed to meet its objectives.
- In smaller businesses, the team might be comprised of just one person tasked with most aspects of the programme, and given the authority to get the job done.

Allocating funds to the GHG management programme can be challenging at the beginning, since the business will not yet have much information about the activities it will undertake and related costs, or about the possibilities for generating cost savings from reduction measures, for example. However, using its high-level goals as a guide, an organization can determine what steps will be required to achieve them, and then gather the information necessary to develop an appropriate budget and timeline. Many aspects of the GHG management programme will not require large upfront investments, and there is a wide range of resources available to help businesses with their inventories and reduction efforts. Proceeding gradually is also an option.

Climate Savers

One objective of GHG management may be participating in a voluntary GHG programme with other businesses. For example, **WWF**'s Climate Savers programme assists leading companies in establishing ambitious targets to voluntarily reduce their greenhouse gas emissions. By 2010, participating companies will have cut carbon emissions by some 14 million tonnes annually – the equivalent of taking more than 3 million cars off the road every year – and saved hundreds of millions of dollars.

SOURCE
www.worldwildlife.org/climate/climatesavers2.html

GETTING STARTED | MEASURING GHGs | REDUCING GHGs | OFFSETTING GHGs | COMMUNICATING | MOVING FORWARD

Helpful resources for getting started

Competitive Advantage on a Warming Planet
Harvard Business Review, March 2007, by Jonathan Lash and Fred Wellington
www.solutionsforglobalwarming.com/docs
HarvardBusReviewonclimatechange-3-07.pdf

Climate Change – A Business Revolution?
by The Carbon Trust
www.carbontrust.co.uk/publications/publicationdetail?productid=CTC740

Consumers, Brands and Climate Change
by The Climate Group
www.theclimategroup.org/assets/resources/research_UK_07.pdf

Brand Value at Risk from Climate Change
by The Carbon Trust
www.carbontrust.co.uk/carbon/PrivateSector/brand_value.htm

**Managing the Risks and Opportunities of Climate Change:
A Practical Toolkit for Corporate Leaders**
by CERES
www.ceres.org//Document.Doc?id=332

Corporate Citizenship: Profiting from a Sustainable Business
by Economist Intelligence Unit
www.eiu.com/site_info.asp?info_name=corporate_citizenship&page=noads&rf=0

The Climate Change Guide
by Canadian Business for Social Responsibility
www.cbsr.ca/sites/default/files/CBSR_ClimateChangeGuide(1).pdf

NETWORKS AND GHG MANAGEMENT PROGRAMMES FOR BUSINESSES

Climate Savers – World Wildlife Fund
www.worldwildlife.org/climate/climatesavers2.html

Climate Leaders – US Environmental Protection Agency
www.epa.gov/climateleaders

Climate Neutral Network – United Nations
www.unep.org/climateneutral/

World Business Council for Sustainable Development
www.wbcsd.org

Environmental Entrepreneurs (E2)
www.e2.org

Investors demand action

In 2009 investors filed a record 68 shareholder resolutions in the United States and Canada to push public companies to address climate change – for example, by measuring and disclosing their carbon footprints, or reporting on the financial and physical risks they face from climate change. 'Investor pressure is prompting more companies to see the value of making their businesses more climate-friendly,' said Mindy S. Lubber, president of **Ceres**, a leading coalition of investors and environmental groups. 'By measuring and lowering the carbon footprint of their operations and products, these companies will have a distinct advantage as the global economy shifts to cleaner energy sources.'

SOURCE
www.ceres.org/Page.aspx?pid=1121

PART 2

Measuring Greenhouse Gas Emissions

Part 2 provides guidance on developing a GHG emissions inventory. It introduces the Greenhouse Gas Protocol, and discusses how to plan an inventory by setting boundaries for the emissions to be measured. It then looks at collecting data for the inventory and calculating GHG emissions, and concludes with ideas for quality control.

O nce a business has determined the goals of its GHG management programme, the next step is often to perform an emissions inventory, which will provide information about the GHG emissions the business is responsible for. This involves identifying the major emission sources of an organization (for example, energy consumption by buildings, and emissions from company vehicles and air travel), and then quantifying those emissions, usually in tonnes of carbon dioxide equivalent (CO_2e).

Although most businesses will have somewhat different emission sources (for example, an office-based company will have a different emissions profile than a retailer; see Diagram 2 at right), there is much common ground, and the same basic accounting approach is used for all types and sizes of businesses.

Why carry out an emissions inventory?

There are several reasons to consider performing an emissions inventory as part of an overall greenhouse gas management plan. First, an emissions inventory highlights the business activities that emit the most GHGs, and helps determine the best opportunities for reductions. Second, the inventory establishes a benchmark for tracking future performance in reducing emissions. Third, the inventory allows a business to assess how it will be affected by potential market and regulatory changes that put a price on GHG emissions. Finally, for businesses that are

DIAGRAM 2:
EXAMPLES OF EMISSIONS
INVENTORY RESULTS FOR
TWO BUSINESSES

LARGE RETAILER

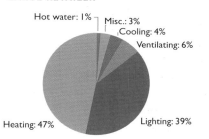

Hot water: 1%
Misc.: 3%
Cooling: 4%
Ventilating: 6%
Heating: 47%
Lighting: 39%

TOTAL: 4,589 TONNES

SMALL OFFICE-BASED COMPANY

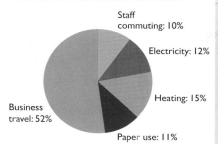

Staff commuting: 10%
Electricity: 12%
Heating: 15%
Paper use: 11%
Business travel: 52%

TOTAL: 232 TONNES

SOURCES
http://oee.nrcan.gc.ca/publications/
infosource/pub/ici/eii/M144_23_2003E/
english/pdf/hosp_eng.pdf
Materials from the David Suzuki
Foundation

pursuing carbon neutral initiatives, the emissions inventory will provide the total amount of emissions that needs to be addressed through direct reductions and the purchase of carbon offsets. To summarize: what gets measured, gets managed.

The Greenhouse Gas Protocol

The Greenhouse Gas Protocol is an internationally accepted accounting method for measuring and reporting greenhouse gas emissions, and is used by companies, governments and non-governmental organizations. It allows the tracking of the six main greenhouse gases that cause climate change (described in the box on page xi). Below is a diagram showing the sources of some of these greenhouse gases.

DIAGRAM 3: BREAKDOWN OF GHG EMISSION SOURCES
ACCORDING TO THE GHG PROTOCOL

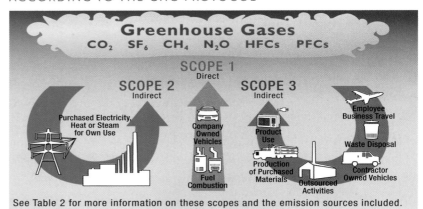

SOURCE: Based on a diagram by New Zealand Business Council for Sustainable Development.

The discussion in this section is adapted from *The Greenhouse Gas Protocol: A Corporate Accounting and Reporting Standard*[5] (referred to in this guide as the 'GHG Protocol'), an excellent resource for businesses and other organizations. It is hoped that this guide will provide an understanding of the core concepts and steps involved, as well as some practical tools to help businesses apply the GHG Protocol to their own emissions inventories.

The primary steps involved in performing an emissions inventory are: establishing an emissions boundary (deciding which emission sources will be included in the inventory), collecting activity data, calculating emissions and ensuring that quality control measures are in place.

Winemakers measure up

Wine industry groups in California, New Zealand, South Africa and Australia have together developed the International Wine Industry Greenhouse Gas Accounting Protocol, based on the original GHG Protocol. Major emission sources for wineries were categorized as follows:

• Scope 1 – Fuel consumption of water heaters, boilers and farm equipment;

• Scope 2 – Purchased electricity, heat and steam; and

• Scope 3 – Supply chain emissions, such as fertilizers and packaging materials, along with emissions from transporting products to market.

The protocol also includes a calculator to help wineries measure their emissions.

SOURCE
www.climatebiz.com/
news/2008/01/28/global-
wine-industry-tackle-carbon-
footprinting

Step 1: Establish an emissions boundary

Before measuring the GHG emissions from a business, it's necessary to decide which emission sources will be included in the inventory. This is called setting the boundaries of the inventory, and is done in two stages: (1) setting the organizational boundary; and (2) setting the operational boundary.

Setting the *organizational boundary* involves looking at the corporate entities that make up the business, and deciding on the basis of the GHG Protocol which of these entities (or perhaps all) will be included in the inventory. Smaller businesses may be a single entity, in which case setting the organizational boundary will be fairly straightforward. Larger businesses may have more complicated forms of organization, such as branches, subsidiaries, joint partnerships and so on; specific guidance for these types of organizations is provided in the GHG Protocol.[6] Once the organizational boundary has been determined, all subsequent inventory activities fall within that boundary.

Next is the *operational boundary*. It defines which GHG emission sources from within the chosen organizational boundary will be included in the inventory. All significant GHG emission sources from the operations within the chosen organizational boundary must be identified, and classified according to the GHG Protocol as 'direct' (owned or controlled by the company) or 'indirect' (owned or controlled by another party). This distinction is designed to avoid double-counting of the same emission sources by different organizations, by assigning responsibility for emission sources based primarily on ownership or control. Direct emission sources are called 'scope 1', and indirect emission sources are divided into 'scope 2' and 'scope 3' (see next page, *Table 2: Emission Scopes According to the GHG Protocol*, for a description of each emission scope).

After classifying its emission sources as scope 1, 2 or 3, a business can decide on its operational boundary, as discussed below. A simple method for mapping the operational boundary is described on pages 14–15.

ISO standards for GHG management

The *ISO 14064 Greenhouse Gases* is a voluntary, three-part series of auditable standards that are compatible with the GHG Protocol. It has been developed and approved by more than 50 countries.

SOURCE
www.csa.ca

Discovering inventories

Discover, a science and technology magazine, recently inventoried the emissions associated with producing almost a million copies of its monthly magazine. Working closely with its suppliers, and using the GHG Protocol, the magazine accounted for emissions from its offices (including employee commuting), forest harvesting, paper production (the largest source by far), printing, disposal and transportation at all stages. The total was 873 tonnes, or 0.95 kilograms of CO_2e for each copy. The company chose to offset these emissions for its 2008 Better Planet special issue.

SOURCE
http://discovermagazine.com/2008/may/21-how-big-is-discover.s-carbon-footprint

---

GETTING STARTED

MEASURING GHGS

REDUCING GHGS

OFFSETTING GHGS

COMMUNICATING

MOVING FORWARD

TABLE 2: EMISSION SCOPES ACCORDING TO THE GHG PROTOCOL

SCOPE 1: DIRECT GHG EMISSIONS (COMPANY-OWNED OR CONTROLLED SOURCES)	SCOPE 2: INDIRECT GHG EMISSIONS (PURCHASED ELECTRICITY, HEAT OR STEAM)	SCOPE 3: INDIRECT GHG EMISSIONS (OTHER SOURCES)
• Generation of heat, steam and electricity • Combustion of fuel in boilers, furnaces or generators • Transportation of materials, goods, products, waste and employees in company-owned or controlled vehicles, planes, ships, etc. • Manufacturing and chemical processing • Fugitive emissions either intentionally or unintentionally released – leaks from equipment joints, for example • Emissions from chemicals (such as HFCs) used in refrigeration and air conditioning equipment	• Purchased electricity • Purchased heat or steam (e.g. through district heating)	• Transportation of goods and materials in vehicles owned or controlled by third parties (e.g. shipping and courier services) • Transportation of people in vehicles owned or controlled by third parties (e.g. business travel, employee commuting and customer travel) • Outsourced activities such as printing, design, etc. • Extraction and production of materials and products (e.g. paper) purchased by the business • End-use and disposal of company products • Consumption of purchased electricity, heat or steam in a leased operation not owned or controlled by the company

SOURCE: *The Greenhouse Gas Protocol: A Corporate Accounting and Reporting Standard* (Revised Edition)

Companies taking responsibility for scope 3 emissions

DHL Express Nordic, a division of a major transportation and logistics company, included scope 3 emissions in its inventory because it found that 94% of its emissions originated from the transport of goods via outsourced transportation firms. Similarly, **IKEA Sweden** decided to include scope 3 emissions from customer vehicle travel to its stores when it became clear that these emissions were large relative to its scope 1 and scope 2 emissions.

SOURCE
The Greenhouse Gas Protocol: A Corporate Accounting and Reporting Standard (Revised Edition)

WHICH EMISSION SOURCES SHOULD BE INCLUDED IN THE INVENTORY?

Determining the operational boundary means deciding which emission sources to include in the inventory. According to the GHG Protocol, at a minimum all scope 1 and scope 2 emissions should be included in the emissions inventory.

Scope 3 emissions are considered optional by the GHG Protocol, but here are some reasons to include significant scope 3 emissions in the inventory:

• **Including scope 3 emissions provides a more realistic accounting of the climate impact of the business.** More and more businesses are taking into account their broader climate impact by looking at scope 3 emissions, which may be large relative to their scope 1 and scope 2 emissions, or crucial to their business activities. For example, a business may include the emissions associated with its supply chain, or with the use of its products by customers. Many businesses already include scope 3 emissions from employee business and commuter travel in their emissions inventories. There is also a growing trend for businesses to consider the full 'life-cycle' emissions of their products, which include all significant scope 1, 2, and 3 emissions associated with these products.

- **The emissions are not being accounted for by any other organization.**
- **Data is readily obtainable.** For example, the emissions associated with the paper used by a business (a scope 3 emission source) can be easily quantified based on the amount of paper used.
- **There are opportunities to engage employees.** Including scope 3 emissions such as employee commuting provides opportunities to engage and empower employees, and also broadens the range of reduction initiatives available.
- **A business may be able to influence the emissions of its suppliers and/or customers.** For example, a business could work with its suppliers to reduce their emissions from shipping – and thereby reduce its own scope 3 emissions.
- **A business wants to show leadership.** A business may wish to include scope 3 emissions in order to take responsibility for its overall climate impact.

CASE STUDY

Salt Spring Coffee Company

DEVELOPING AN EMISSIONS INVENTORY

Salt Spring Coffee Company is a small coffee roaster located on Salt Spring Island, in British Columbia, Canada. A pioneer in its industry, Salt Spring Coffee Company sells coffees that are 100% certified organic, fair trade and shade tree-grown.

In 2007, Salt Spring coffee company launched its Carbon Cool initiative. With this initiative Salt Spring aimed to analyse its GHG emissions and find ways to reduce its carbon footprint, to use its programme as a way to educate its customers about global warming, and also to further differentiate itself in a competitive market.

Salt Spring measured all of the major scope 1 and 2 emission sources associated with its own production and distribution, such as roasting the beans, transporting the coffee to stores in its vehicles, and electricity use. But it also included several key scope 3 emission sources in its operational boundary, such as air travel and the shipping of green beans from their port of origin.

Breakdown of 2007 GHG emissions (in metric tonnes CO_2e)

EMISSION SOURCE	SCOPE	AMOUNT
Propane use in roasting	1	131.6
Company vehicle use	1	36.2
Heating	1	9.6
Refrigerants	1	1.0
Electricity	2	7.1
Air travel	3	31.4
Green bean shipping	3	31.2
Ferry travel	3	15.3
TOTAL		263.4

To ensure the quality and accuracy of its inventory, Salt Spring had consultants review its GHG calculations, which were confirmed to be 263.4 tonnes in 2007. The company calculates that this total translates into approximately 838 grams of CO_2e per bag of coffee, for which it purchases offsets annually.

FOR MORE INFORMATION
www.saltspringcoffee.com

Following the steps
on these two pages, a
business can identify its
emission sources; classify
them as scope 1, 2 or 3;
and draw an operational
boundary for its emissions
inventory.

Mapping the Operational Boundary

One way to determine the operational boundary for a business is to draw a simple map all of its physical sites and related emission sources, and determine whether they are scope 1, 2 or 3 by following the steps below. It may be helpful to walk around the facilities to collect some of the information required. See the diagram on the next page for an example of a completed map.

NOTE: businesses should first define their 'organizational' boundary, i.e. if they have branches, subsidiaries or partnerships they should determine which of these will be included in their inventory.

EQUIPMENT NEEDED: paper; black, red, green and blue markers

1 Identify physical sites

To begin, identify all of the physical sites within the organizational boundary. Examples include head and branch offices, garages, warehouses, etc. *Using a black marker, draw a building to represent each site and then label.*

2 Map heating and electricity for all physical sites

A For each site indicate whether there is heating or electricity generated by equipment owned by the business. *Draw a red arrow to these sites and label it.*

B For sites that use purchased electricity, heat or steam, *draw a green arrow and label it.*

C In some cases, such as leased spaces, the business may not have ownership or operational control of the site, particularly if it is not the sole tenant (see the *Glossary* for a definition of *control*). If there is no ownership or control, *indicate any electricity, heat or steam that is purchased or included in the lease with blue arrows, and label.*

3 Map other GHG emission sources

When mapping each of the emission sources below, draw the boxes below the physical site they are associated with. In the case of emission sources that are common to more than one site, either draw separate boxes for each site or make a note on the box that it applies to more than one site.

A TRANSPORTATION

- Identify all company-owned vehicles, such as company cars and delivery fleets. *Draw a red box for each group and label.*

- Identify all transportation of people and freight in vehicles not owned by the business. Some examples include: transporting goods to and from the business by truck, rail or air; business travel by employees; and employee commuting. *Draw a blue box for each source and label.*

B OTHER ON-SITE GHG EMISSIONS

- Identify all other on-site GHG emission sources for the business, such as the fugitive emissions of chemicals used in refrigeration or manufacturing. *Draw a red box for each source and label.*

C MATERIAL INPUTS

- Identify material inputs (e.g. packaging, paper and other supplies) sourced from outside the business. *Draw a blue box for each input or group of inputs and label.*

D OUTSOURCED SERVICES

- Identify outsourced services, such as cleaning services. *Draw a blue box for each service and label.*

DIAGRAM 4: EMISSION SOURCES AND OPERATIONAL
BOUNDARY FOR GOURMET FROZEN FOODS, INC.

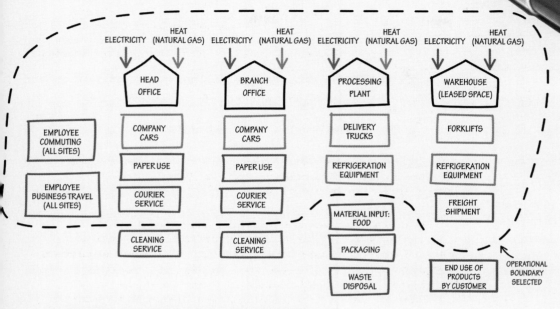

E END-USE OF PRODUCTS BY CUSTOMERS
- Identify products sold by the business – their end-use will produce emissions. *Draw a blue box for each product or group of products and label.*

F WASTE DISPOSAL
- Identify waste produced by the business, as this will produce emissions when landfilled. *Draw a blue box for waste and label it.*

G EXTERNAL EVENTS (CONFERENCES, RETREATS, ETC.)
- Identify any external events hosted by the business. *Draw a blue box for each event and label.*

4 Draw the operational boundary

   A Include all red and green items. These are scope 1 (red) and scope 2 (green) emissions according to the GHG Protocol. (See Table 2 on page 12 for a description of scope 1, 2 and 3 emissions.)

   B Decide which scope 3 emissions to include – these are all the blue items. For more guidance, see *Which emission sources should be included in the inventory?* on page 12.

   C Use a black marker and draw a dotted line around emissions that will be included in the inventory. This is the operational boundary.

Once the operational boundary has been determined, the next step is to collect emissions activity data for all of the emission sources that have been included. This is discussed on page 16, and Table 3 gives a list of common emission sources and where to find data.

    Businesses should also note that they may wish to expand their operational boundary over time, and include more emission sources in their inventory. Another possibility is to work with the organizations responsible for the emissions (such as suppliers) and encourage them to make reductions.

SOURCE: Adapted from materials developed by the Pembina Institute, www.pembina.org

GETTING STARTED

MEASURING GHGS

REDUCING GHGs

OFFSETTING GHGs

COMMUNICATING

MOVING FORWARD

Step 2: Collect activity data

Once the emissions boundary has been drawn, the next step is to collect the relevant information about each emission source within the boundary. This is referred to as *activity data*. Activity data is simply a measurement of the activities that generate emissions, and will be found in standard units such as kilometres driven, litres of fuel used, and kilowatt hours.

Collecting activity data is often the most time-consuming part of developing a GHG inventory, simply because company records might not have captured this data in a systematic way.

Because the accuracy of the emissions inventory is only as good as the activity data it is based on, it is important that the data be collected carefully. If data is not available for some emission sources, businesses may need to rely on estimates, particularly in the first years of carrying out an inventory. As expertise and information are acquired within the organization, the precision of the inventory will increase.

In order to simplify and standardize the data collection process, many companies design an information management system for data collection. This involves identifying which data needs to be collected, where the data will be obtained, who will be responsible for collection, and how the data will be managed and stored over time. For smaller companies, this is a relatively straightforward process, but larger companies with more than one facility may require a more sophisticated system, such as an online emissions database.

Table 3 provides some examples of common emission sources, and typical places to find corresponding activity data.

Measuring the footprint of a glass of orange juice

PepsiCo decided to measure the carbon footprint of its Tropicana® orange juice. It looked at emissions from growing the oranges, processing them into juice, distribution, packaging, end-use and disposal. The results were surprising: the largest source of emissions was from growing the oranges. Citrus groves typically rely on nitrogen fertilizer made from natural gas, which can lead to nitrous oxide, a potent greenhouse gas. PepsiCo found that of the 1.7 kg carbon footprint for each 1.89 litre carton of orange juice, over a third of the emissions came from fertilizer use. PepsiCo plans to work with its growers and with researchers to find ways to grow oranges that are less greenhouse gas intensive.

SOURCES
www.nytimes.com/2009/01/22/
business/22pepsi.html
www.tropicana.com/pdf/
carbonFootprint.pdf

Streamlining the inventory process

Over time, creative ways can be found to make the data collection system faster and more efficient. For example, if employees fly frequently, one option is to work with the accounting department to devise a system to capture flight data as trips are booked. Another option is to arrange to have the company travel agent keep a log of air travel for the business including distances for each flight.

TABLE 3: COMMON EMISSION SOURCES AND WHERE TO FIND ACTIVITY DATA

EMISSION SOURCE	WHERE TO FIND ACTIVITY DATA	TYPICAL UNITS / DATA TYPE
Purchased electricity	Utility bills; online customer accounts.	kWh, MWh
Purchased heat (e.g. district heating)	Utility bills; online customer accounts.	GJ, BTUs, therms, MWh, lbs of steam
On-site heat generation (e.g. furnaces)	Utility bills; fuel purchase records and invoices; storage tank logs; online customer accounts.	Litres, gallons, GJ, m³, cubic ft, kg, lbs, BTUs
Company-owned vehicles	Fuel purchase records; fuel receipts; fuel tank logs. If fuel consumption data is not available, kilometres travelled (trip records, odometer readings, maintenance records) and vehicle make and model year are a second-best option when used with online vehicle emission calculators.	fuel type and amount (litres/gallons), or km/ miles travelled and vehicle make/model year
Business travel (vehicle)	Accounting receipts; expense claims.	fuel type and amount (litres/gallons), or km/ miles travelled and vehicle make/model year
Business travel (air)	Online calculators can be used to find distances for flights.	km/miles flown, or possibly fuel used (in the case of charter flights)
Employee commuting	Many employers use surveys to collect data about this – free online survey providers can be useful.	km/miles travelled and mode of transportation, or fuel type and amount consumed
Freight transport	Shipping invoices; delivery invoices. Shipping and delivery companies may need to be contacted to obtain the information required, i.e. the weight shipped and the distance travelled.	kg/lbs/tonnes and km/ miles transported, and mode of transport (truck, rail, air, ship)
Leased space	Lessees might not receive bills for electricity or heating/cooling charges. In this case average calculations can be done based on the size of the space and the length of time it is used.	m² or sq. ft and number of days
Material inputs	Receipts for purchases; suppliers; life-cycle analysis calculators.	Varies
Fugitive emissions (air conditioning/refrigeration equipment, pipelines, etc.)	Industry and government publications; equipment specifications.	Varies
Outsourced services	Request this information from contractors or suppliers, or work with them to obtain it.	Varies

Step 3: Calculate GHG emissions

Once all relevant activity data has been collected, the next step in creating an inventory is to do the emissions calculations. The basic formula is:

(activity data) x *(emission factor) = GHG emissions*

Emission factors are used to convert activity data from a business into GHG emissions values (usually in kilograms or tonnes of carbon dioxide equivalent, CO_2e). For example, using the emission factor for long-haul air travel, it is possible to calculate the emissions produced by one passenger travelling 1000 km by airplane:

1000 km x *0.11 kg CO_2/km = 0.11 tonnes CO_2*

Other emission factors allow the calculation of CO_2e produced by using electricity or driving a car. Most often, businesses will use calculation tools, either in an online or spreadsheet format, to calculate their emissions. These tools are easy to use (since they use built-in emission factors), save time, and can reduce the chance of errors. *Helpful resources for measuring GHG emissions* (page 19) includes links to some of these tools, along with commonly used emission factors.

Tips for creating an emissions inventory

- An inventory is an iterative process, not a one-time project. Expect to improve accuracy and expand scope over time, rather than aiming for perfection the first time around. However, make sure that opportunities for improvement are captured, e.g. if data from fuel consumption from company cars is missing, determine a way to collect that information for the next inventory and put it in place.
- Similarly, a business might start with a relatively basic inventory, in terms of scope, and then expand it in succeeding years. For example, in its first-year inventory, a business could include all scope 1 (e.g. company fleet and building energy use) and scope 2 emissions (e.g. purchased electricity), and possibly some relevant scope 3 sources (e.g. business air travel). Additional scope 3 emission sources could be added in subsequent years.
- When starting, it may be useful to focus on what can readily be measured, and then implement some quick reduction measures in those areas. This will help create momentum by demonstrating that the programme delivers results.
- The first inventory might not be the ideal baseline, i.e. the benchmark for comparing future inventories, if it is less than complete or known to have problems with accuracy. See *Step 1: Set a reduction target* in Part 3 for more information.

Step 4: Quality control of the emissions inventory

Quality control of the emissions inventory is important. The information provided by the inventory may have many different uses, each of which requires reliable information: identifying areas for targeting reductions; evaluating progress in reducing emissions in future years; assessing the exposure of an organization to factors like carbon pricing; meeting regulatory requirements; making public claims related to the carbon footprint of the business (e.g. in an annual report); or calculating the number of offsets required for a carbon neutral initiative.

It is fairly easy to make simple mistakes that, when multiplied at the calculation stage, result in an emissions total that is significantly higher or lower than it should be – for example, by using incorrect units (e.g. miles instead of kms), using incorrect emission factors, or through simple data entry errors. For this reason, if a business is preparing its own calculations it is useful to have a second employee review the calculations and data entry sheets.

It can also be reassuring for organizations that are preparing their own inventories to have a reputable consultant review their emissions calculations. Larger organizations might wish to have a more formal third-party verification statement by an auditor. Most major accounting firms now provide these services, as do a number of smaller consulting operations.

Helpful resources
for measuring GHG emissions

The Greenhouse Gas Protocol: A Corporate Accounting and Reporting Standard *by World Resources Institute*
www.ghgprotocol.org/files/downloads/Publications/ghg-protocol-revised.pdf

Hot Climate, Cool Commerce: A Service Sector Guide to Greenhouse Gas Management *by World Resources Institute*
http://pdf.wri.org/hotclimatecoolcommerce.pdf

Working 9 to 5 on Climate Change: An Office Guide
by World Resources Institute
www.wri.org/publication/working-9-5-climate-change-office-guide#

Making Advances in Carbon Management: Best Practice from the Carbon Information Leaders
by Carbon Disclosure Project and IBM
www-05.ibm.com/innovation/uk/green/pdf/joint_cdp_and_ibm_study.pdf

---

Software to help inventory GHG emissions from energy use

Johnson Controls, a specialist in climate control systems, developed a software system for its clients that processes data from their utility bills to help them measure and report GHG emissions. It also enters into performance contracts with its clients, under which it commits to reduce GHG emissions at client facilities using the software.

SOURCE
Citigroup Equity Research: Global Thematic Investing, Climatic Consequences, 19 January 2007 p. 93.

GETTING STARTED

MEASURING GHGs

REDUCING GHGs

OFFSETTING GHGs

COMMUNICATING

MOVING FORWARD

GHG EMISSION CALCULATION TOOLS

GHG Protocol Calculation Tools
www.ghgprotocol.org/

Environmental Defense Paper Emissions Calculator
www.environmentaldefense.org/papercalculator/

Carbon Calculators for Businesses and Other Organizations
www.davidsuzuki.org/Climate_Change/What_You_Can_Do/business_carbon_calculators.asp

OpenEco.org
www.openeco.com

CASE STUDY

David Suzuki Foundation
MEASURING EMISSIONS FROM AN OFFICE-BASED ORGANIZATION

As an environmental organization working on sustainability and climate solutions, the David Suzuki Foundation is committed to reducing its own climate impact. Since 2003, staff have carried out an annual inventory of the greenhouse gases generated by the Foundation's activities.

Using resources from the GHG Protocol website, four major emission sources were originally identified and included in the inventory: electricity use, staff commuting, paper use, and air travel by staff. In 2007 an additional source – events – was added to reflect the fact that several public events were organized by the Foundation that year. For the fiscal year 2007–2008, greenhouse gas emissions from these sources totalled 183.6 tonnes.

Over the years, the Foundation has been able to simplify the process of completing its annual inventory. For example, it has worked with its key suppliers to find ways to collect activity data. The Foundation's printing company sends an annual tally (in kilograms, plus % recycled content) of all paper used by the Foundation, and its travel agent sends a monthly spreadsheet with all air travel, including distances in kilometres for each flight taken. Data from staff commuting is captured at the same time the Foundation participates in the annual Commuter Challenge event (see box on page 40).

The Foundation continues to look for ways to reduce its emissions. Some examples include: choosing an energy efficient photocopier; installing timers on all major appliances so they shut off at the end of the day; and using 100% post-consumer recycled paper in its photocopier, printers and most publications. The Foundation also recently installed video-conferencing equipment for its offices that will help to reduce emissions from air travel, as well as save money on air fares.

To compensate for emissions that remain after reduction efforts, the Foundation purchases high quality, Gold Standard carbon offsets.

FOR MORE INFORMATION
www.davidsuzuki.org/Climate_Change/What_You_Can_Do/carbon_neutral_office.asp

PART 3

Reducing Greenhouse Gas Emissions

Part 3 discusses how to reduce greenhouse gas emissions from a business. It covers setting a reduction target, identifying and selecting reduction opportunities, tracking reductions and cost savings over time, and concludes with a summary of typical reduction opportunities for businesses.

Reducing emissions is the single most important part of an organization's greenhouse gas management programme. Without reductions, the problem of climate change cannot be solved.

By reducing its own GHG emissions, a business can not only reduce its own climate impact, but also potentially realize some other important benefits, such as cost savings.

The main steps involved in reducing emissions are:

1. **Set** a reduction target.
2. **Identify** opportunities for reducing GHG emissions.
3. **Assess**, select and implement the emission reduction measures.
4. **Track** reductions and cost savings on a regular basis.
5. **Continue** to make reductions, and look for new reduction opportunities.

Each of these steps is discussed in more detail below.

Small changes add up to big GHG reductions

HP, a global IT manufacturer and supplier, estimated that its redesigned print cartridge packaging for North America would reduce greenhouse gas emissions from shipping by an estimated 16,000 tonnes in 2007 – the equivalent of taking more than 3,000 cars off the road for one year.

SOURCE
http://h41131.www4.hp.com/ca/en/pr/02082007a.html

GETTING STARTED

MEASURING GHGs

REDUCING GHGs

OFFSETTING GHGs

COMMUNICATING

MOVING FORWARD

Step 1: Set a reduction target

A business can often drive GHG reductions more effectively by setting an emission reduction target. For example:

- A target provides a business with a concrete goal around which reduction efforts can be planned, and performance assessed.
- A target is an excellent way for an organization to communicate its commitment to reduce its climate impact to employees and other stakeholders.
- Having departmental or unit sub-targets can help to establish accountability. Involving employees in the process of setting targets can also be a way to ensure their buy-in.
- Businesses that are participating in government or voluntary GHG programmes may be required to set a reduction target.

As with other aspects of GHG management programmes, there are many options with respect to targets. For example, a company-wide target could be set for all emission sources and business units: for example, 10% below 2009 emissions by 2012. Or a business can set diverse targets across different business units or emission sources that all contribute to the overall company target: for example, reducing electricity use by 20% and reducing air travel by 15%. A business whose emissions result primarily from transportation might set a target to reduce just those emissions.

Whatever target is chosen, it is important that it be a challenging and inspirational goal that will motivate employees to think creatively, and help propel operations beyond 'business as usual'.

CHOOSING A BASE YEAR FOR THE REDUCTION TARGET

In order to set a reduction target, a business needs to first select a base year, which will be a reference year against which to assess its progress in meeting its reduction target in future years. It is important to choose a base year for which reliable, accurate and comprehensive emissions data is available. The base year may be the first year that an emissions inventory is completed, or, alternatively, the first year that a business decides its emissions inventory is sufficiently accurate and complete. A specific base year may also be prescribed by government or voluntary GHG programmes.[7]

Green refrigeration equipment saves money and GHGs

GreenChill is a voluntary programme of the US EPA that is helping supermarkets to reduce greenhouse gas and ozone-depleting emissions. Twenty-eight partners, including several large supermarket chains, are preventing leakage of refrigerants and/or installing more efficient refrigeration systems. New technologies tend to be more energy efficient, need less maintenance, improve the shelf life of food, and produce fewer GHG emissions. So far, the programme has saved nearly $13 million in operating costs, and avoided emissions of 2.5 million tonnes of CO_2e.

SOURCE
www.greenbiz.com/news/2008/06/10/greenchill-members-save-13m

HOW DEEP SHOULD THE REDUCTION TARGET BE?

How deep the reduction target is depends on several factors:

- **Level of opportunity for GHG reductions.** A business that has not previously made efforts to reduce its emissions can usually meet more aggressive targets, since there will be more opportunities for reductions.
- **Timing.** A business that is due to make large investments in new equipment or facilities may have significant reduction opportunities.
- **Flexibility.** A business that is able to adjust its maximum payback period for investments in energy efficiency or other GHG reduction measures (e.g. from three years to five years or more) will have more cost-effective options available to it.
- **The type of emission sources** in the business's inventory, because some sources are easier to reduce than others.
- **Expected future growth of the business**, which could be accompanied by growth in emissions.

Absolute vs. intensity targets

It is important to understand the distinction between absolute targets and intensity targets. Absolute targets are concrete emission reduction goals for the entire organization, or for specific emission sources, and do not consider other factors, such as the company's growth. An example of an absolute target would be the reduction in total greenhouse gas emissions by 20% below 2000 levels by 2010. Absolute targets are needed to effectively address climate change, because, if met, they ensure that the overall amount of emissions that enters the atmosphere is reduced. Businesses that want to reduce their climate impact should be setting absolute targets for emission reductions.

Intensity targets are relative to some measure of business activity, such as company growth, or units of production. An example of an intensity target would be the reduction of CO_2 per unit of production by 10% between 2000 and 2008. Meeting intensity targets will not necessarily result in a reduction in a company's overall emissions. For example, if a company's production increases, its overall emissions may increase as well, even if the intensity target is met. Intensity targets can be useful to measure progress within a company in reducing the energy intensity of various activities, but they are not a replacement for absolute reduction targets.

Working with suppliers to reduce GHG emissions
Wal-Mart, the world's largest retailer and private employer, in partnership with the Carbon Disclosure Project, is asking a group of its suppliers to measure and report the energy used to make and distribute their products. A pilot scheme will involve about 30 interested suppliers in seven product categories: soda, beer, soap, toothpaste, DVDs, vacuum cleaners and milk. The goal is to find ways to reduce Wal-Mart's indirect climate impact, and it is expected that this will also lead to cost savings for the company.

SOURCE
www.greenbiz.com/news/2007/09/24/wal-marts-newest-green-goal-cleaner-supply-chains

GETTING STARTED

MEASURING GHGs

REDUCING GHGs

OFFSETTING GHGs

COMMUNICATING

MOVING FORWARD

As simple as fixing a leak
At **Catalyst Paper Corporation,**
one of North America's largest
producers of mechanical printing
papers, employees are identifying
dozens of simple ideas to save
energy. One mill calculated that
fixing an air leak in a quarter-inch
pipe would save $6,000/year in
wasted energy, while a one-inch
air hose left running costs the mill
$54,000/year.

SOURCE
www.worldwildlife.org/climate/
climatesavers2.html

Step 2: Identify opportunities for reducing GHG emissions

A good place to start when deciding where to make reductions is with the completed emissions inventory, as it will often reveal the best opportunities for reductions. In many cases the largest emission sources will offer the most potential for making reductions. Employees can also be instrumental in helping identify reduction opportunities.

There are many different ways for businesses to achieve reductions in GHG emissions. Later in Part 3 is a description of general categories of reduction types, and some common examples of specific reduction opportunities. This list is far from exhaustive, but these categories are a good starting point for many organizations. Businesses may also find it helpful to work with a consultant to identify cost-effective reduction opportunities through an energy audit or other assessment measures.

**Look for the
ENERGY STAR®**
Using ENERGY STAR qualified
equipment can save over $3,000
per year for an office of 200
employees. A current list of
ENERGY STAR qualified products
in the US can be found at www.
energystar.gov, and in Canada at
www.energystar.gc.ca.

SOURCE
NRCAN. The ENERGY STAR® is
administered and promoted in Canada by
Natural Resources Canada and is registered
in Canada by the United States Environmental
Protection Agency.

Although it can be tempting to begin with low-cost, low-tech solutions, and over time implement solutions that require more investment, this practice – while yielding immediate savings – tends to leave a business with more expensive measures that become increasingly difficult to justify. One way to avoid this is to bundle a combination of short- and longer-term payback initiatives, which will allow a business to realize some immediate energy savings yet still tackle longer payback items. Another option is to earmark and reinvest the savings achieved by each measure into future reduction efforts.

Most businesses begin looking for reduction opportunities within their own operations, and then proceed to engage with suppliers and others to deal with upstream or downstream emissions.

GETTING STARTED

MEASURING GHGs

REDUCING GHGs

OFFSETTING GHGs

COMMUNICATING

MOVING FORWARD

CASE STUDY

Walkers

USING A GHG INVENTORY TO COMPARE REDUCTION OPPORTUNITIES

Walkers is the UK's largest snack food manufacturer with brands such as Walkers Crisps and Walkers Sensations. Walkers estimates that 11 million people eat one of its products every day.

After deciding to manage its greenhouse gas emissions, Walkers worked to measure and reduce the impact of its own manufacturing processes, and has reduced its energy use by 30% below 2000 levels. It also began sourcing all of its potatoes within the UK to reduce transportation emissions.

However, Walkers recognized that its own operations were just one link in the chain of production for its snack foods. In cooperation with the Carbon Trust, a body set up by the UK government to work with businesses to reduce their carbon emissions, Walkers set out to inventory the emissions created over the entire life cycle of its products. This included emissions associated with growing the potatoes and other ingredients, materials used in packaging, manufacturing, delivery to stores, and disposal and recycling of packaging and waste.

Looking closely at results of the inventory yielded some important insights. For example, it turned out that Walkers' pricing structure for potatoes was actually encouraging greater greenhouse gas emissions. At the time of its inventory, Walkers was purchasing its potatoes from farmers by weight. The result was an incentive for farmers to sell heavier potatoes, which they did by storing their potatoes prior to sale in artificially humidified warehouses. This meant that farmers were using energy – and creating extra emissions – in order to increase the water content and weight of their potatoes. At the manufacturing stage, Walkers was using even more energy to dry and fry the moister potatoes.

The solution? Walkers adjusted its pricing structure to favour potatoes with lower water content. This allowed Walkers to save energy at the manufacturing stage, and potato farmers to save energy by not having to humidify their warehouses. This simple change had the potential to save up to 9,200 tonnes of greenhouse gas emissions and £1.2m annually across the supply chain. Conducting an inventory allowed Walkers to identify this and other important reduction opportunities, which often saved costs at the same time.

Not only did Walkers engage with its suppliers to find reduction opportunities, but it also brought the results of its emissions inventory to its customers. It became the first company to introduce the Carbon Trust's Carbon Reduction Label on its packaging, which shows the carbon footprint of one bag of crisps (80 grams of CO_2e). The label also carries Walkers' commitment to reduce its emissions, and informs customers about the impact of their own choices.

FOR MORE INFORMATION
www.walkers.co.uk

Step 3: Assess, select and implement GHG reduction measures

Deciding which reduction opportunities to pursue will require an assessment of their relative merits. While the ability of a particular measure to achieve GHG reductions and save money is obviously very important, all of the considerations listed below need to be weighed carefully.[8]

- Ability of the measure to reduce GHG emissions of the business.
- Cost to implement (capital costs, installation costs, operating and maintenance costs, and associated staff time and costs).
- Time to implement.
- Payback schedule (how many years it will take for a measure to pay for itself in energy savings).
- Net return on investment (ROI).[9]
- Committed internal champions for the measures.
- Existing momentum from complementary initiatives.
- Degree of cooperation required from company departments or offices.
- Opportunities to have external support or collaboration.
- Visibility and contribution to brand worth.
- Collateral benefits to the company, the environment and the community.
- Barriers to implementation.

Once the emission reduction measures are selected, their implementation can be incorporated into a business's operating and maintenance schedules.

Businesses have many options when it comes to financing emission reductions, including:

- **Internal financing.** In addition to budgeting for reduction initiatives, businesses can use savings from previous reduction initiatives to finance new ones.
- **Financial lenders.** Some banks now offer green business loans at special rates for energy efficiency projects, and most provide regular loans if an attractive return on investment can be demonstrated.
- **Government programmes.** Many government incentive programmes exist for businesses, often providing specialized assistance, including programmes to help reduce energy use or purchase hybrid vehicles.
- **Utilities.** Many utilities offer incentives to businesses to reduce energy use and may provide energy audit services as well.

- **Energy service contractors.** Businesses that lack the funds to invest in more capital-intensive energy efficiency projects may be able to work with energy service contractors (ESCOs). An ESCO contracts with a business to install energy efficiency technologies at the business's premises, at the ESCO's expense. The ESCO is repaid over time through a percentage of the energy savings that result from the technology installed. ESCOs can be a viable option for many businesses because of their experience with energy savings initiatives, and the fact that no upfront expenditures are necessary.

  For more information, see the *Helpful resources* at the end of Part 3.

Step 4: Track GHG reductions and cost savings on a regular basis

Tracking allows a company to monitor the success of its reduction efforts. For most businesses it will make sense to track reductions and any related cost savings on an annual basis, although there may be specific projects where shorter- or longer-term monitoring is desirable. Because a company's emissions inventory is also usually done annually, it can be designed to incorporate more detailed information about reduction efforts so that tracking emissions (and financial savings) over time is possible. For example, the annual total electricity emissions could be broken down by facility in the inventory to see where reduction efforts have been effective, and where more effort is needed. Where possible, businesses may wish to track this data on a monthly or quarterly basis to facilitate year-end reporting and account for seasonal variations.

Step 5: Continue to make GHG reductions

Like the entire GHG management programme, making reductions is an iterative process. Each year, there will be further opportunities to progressively reduce the company's emissions. Ideally, successful reductions in one area will catalyse reductions in another. There are also many ways to encourage new ideas, including staff brainstorming sessions, and a suggestion box that rewards good ideas. Part 6, *Moving Forward*, provides ideas about how to incorporate GHG management, including reducing emissions, into the structure and culture of a business, as this can advance reduction efforts considerably.

Reduction targets should be assessed annually, and if they are going to be met ahead of schedule, an organization may wish to set a new, more ambitious target. On the other hand, a business may find that one or more of its reduction measures has not been as successful as desired. These measures can be evaluated and the results used to refine the process of assessing future reduction opportunities and targets.

Tracking the savings

IBM operates in more than 170 countries and has approximately 400,000 employees. The company regularly monitors its energy use and greenhouse gas emissions, enabling it to track its cost savings from reductions as well. In 2008, the company's energy projects saved 235 million kWh of electricity and 6.3 million gallons of fuel – reducing emissions by 215,000 tonnes of CO_2, and saving $32.3 million in energy expenses. Overall, the company calculates that, between 1990 and 2008, it has avoided 3.3 million tonnes of CO_2 as a result of its energy projects and has saved a total of $343 million in energy expenses.

SOURCE
www.ibm.com

Examples of GHG emission reduction opportunities

Some of the more common emission reduction opportunities for businesses are outlined below. The list is far from exhaustive, but provides an idea of the possibilities that exist. For convenience, they are divided into:

- Energy use.
- Transportation.
- Renewable energy sources.
- Operational efficiency.
- Material inputs.
- Upstream GHG reductions from suppliers and contractors.
- Downstream reductions (e.g. from customers).

A. ENERGY USE

Significant GHG emissions are associated with energy use, including from electricity, and from fuels used for heating, cooling and industrial processes. There are many ways that businesses of all sizes can reduce their energy use, and also their GHG emissions and energy costs.

There are two broad categories of measures to save energy: (1) energy conservation measures, which usually incur little cost, and (2) energy efficiency investments, which range from minor costs to more substantial outlays.

1. Energy conservation measures are usually good housekeeping practices that involve minor changes in employee work practices, incur little or no cost, can usually be implemented quickly, and generate immediate and ongoing savings. Most businesses can easily take advantage of some of these measures, and the GHG reductions can be significant.

It should be noted that many energy conservation measures rely on changes in employee behaviour, and thus require continual reinforcement in order to be effective, especially when large numbers of employees are involved.

Canada's first solar-powered laundromat

Located in Toronto, **Beach Solar Laundromat** is Canada's first solar-powered laundromat. Owner Alex Winch installed eight solar hot water panels and reduced natural gas consumption by 30%. He also switched all of his T12 lighting to T8s. Revenues grew by 160% over 18 months as new customers chose to use the laundromat due to its environmentally friendly energy initiatives.

SOURCE
www.cleanairfoundation.org/coolshops

A good tip for saving energy

The UK Environmental Manager for **IKEA** suggests an easy way to identify opportunities for energy savings: come in at 5:00 a.m., and see what's been left on.

SOURCE
www.ethicalcorp.com/content.asp?ContentID=5339

DIAGRAM 5: EXAMPLES OF GHG EMISSION REDUCTION OPPORTUNITIES

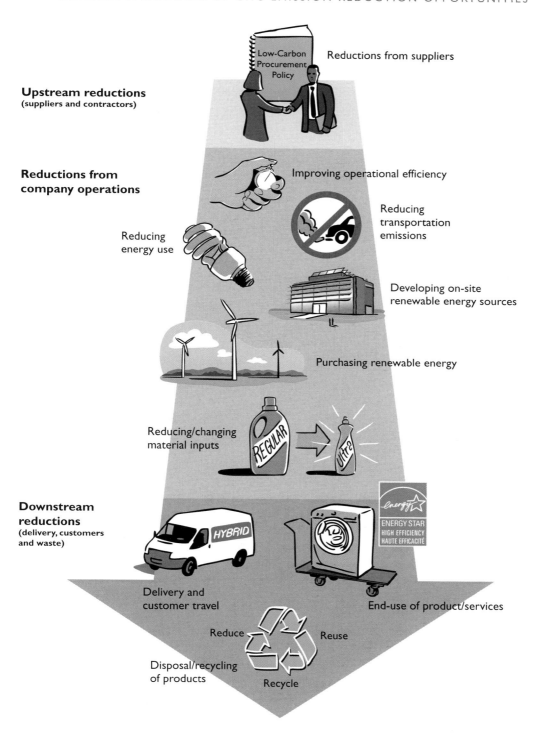

Reductions from suppliers

Upstream reductions
(suppliers and contractors)

**Reductions from
company operations**

Improving operational efficiency

Reducing
transportation
emissions

Reducing
energy use

Developing on-site
renewable energy sources

Purchasing renewable energy

Reducing/changing
material inputs

**Downstream
reductions**
(delivery, customers
and waste)

Delivery and
customer travel

End-use of product/services

Reduce   Reuse

Disposal/recycling
of products

Recycle

GETTING STARTED

MEASURING GHGs

REDUCING GHGs

OFFSETTING GHGs

COMMUNICATING

MOVING FORWARD

Providing training may be an effective option where there is a building operator involved, for example. Alternatively, some of these changes can also be achieved through devices (like room occupancy sensors, timers, or programmable thermostats) that do not rely on employees.

Some examples of energy conservation measures include:

- **Lighting.** Turn off lights at night and when rooms or areas are not being used, or install occupancy sensors that automatically switch off lights. Maximizing the use of natural light will also reduce the amount of energy consumed, and can be accomplished through the use of windows, skylights, or other design features.
- **Heating and cooling.** Keep exterior doors closed; close blinds to reduce heat build-up from direct sunlight; adjust thermostats at times when buildings are not in use (or use a programmable thermostat); make sure all vents are unobstructed; change furnace filters and service HVAC (heating, ventilation and air conditioning) systems regularly; use windows for ventilation.
- **Computers and other office equipment.** Turn off computers, monitors, photocopiers, printers and other office equipment at the end of the day, and programme the equipment to go into 'sleep' mode when not in use. Timers can also be used for equipment like printers and photocopiers so that they are automatically turned off at the end of the day.

2. Energy efficiency investments are measures that require some investment and time to implement, but will usually have attractive payback periods. They range from relatively inexpensive measures, such as installing occupancy sensors, to more advanced initiatives, such as designing new green buildings. Businesses can save money from these measures, but also benefit in other ways. For example, natural light has been shown to improve employee productivity, and measures to improve HVAC systems can improve air quality and reduce absenteeism.

To identify potential energy conservation measures, ask the following questions:

- Do the lights or equipment need to be on as long as they are?
- Can the operating temperature be reduced?
- Can smaller, more efficient equipment be installed?
- Can insulation be added?
- Can leaks in windows and doors be sealed, or is replacement necessary?

SOURCE
www.climatechangeconnection.org/ Resources/documents/Business_Guide. pdf

---

Saving money through energy management

Ken Harvey, group CIO of **HSBC**, one of the world's largest banks, notes the impact of rising fuel prices: 'As the price of oil has hit the roof, the cost of powering datacentres has become something IT directors can no longer ignore. Green is not just a nicety – careful power management will actually help businesses save money.'

SOURCE
www.computerweekly.com

Computing the savings
Every computer turned off when not in use case can save up to $70 per year in energy costs.

SOURCE
NRCAN

A business can time the implementation of these measures to coincide with the regular replacement of equipment or the expansion of operations. Businesses that lease their premises will likely have less flexibility with respect to building energy infrastructure, but they can still make energy efficiency a priority when they are negotiating or renegotiating leases.

Some examples of energy efficiency investments include:

- **Lighting.** Replace incandescent light bulbs with compact fluorescents. For buildings that already use fluorescent lighting, replace T12 lamps and ballasts with higher efficiency T8 or T5 lamps and electronic ballasts, which use at least 40% less energy. LED lighting is also quickly becoming an option for many lighting needs, and uses much less energy and requires less maintenance.

- **Computers and other office equipment**. When it is time to replace old equipment and computers, purchase ENERGY STAR® qualified models, which use less energy and also produce less heat when used. Data centres are also becoming targets for energy reductions, because energy consumption and cooling requirements of computer servers have increased dramatically.

- **Heating and cooling.** Install improved HVAC controls, apply caulking around windows and building entrances and exits, improve building insulation, install more efficient heating and cooling equipment and water heaters.

- **Building recommissioning.** This involves having an expert 'tune up' a building's mechanical equipment, including the HVAC, controls and electrical systems, which are analysed for proper operation and then optimized. In some cases, recommissioning can help avoid the need to install new or additional equipment.[10]

- **Cogeneration.** Also known as 'combined-heat-and-power', cogeneration can be used to capture heat from on-site sources such as electricity generation or industrial processes and use it for heating or other purposes.

Four-day work week cuts emissions

Moving to a four-day work week has turned out to be as popular with some employers as it is with their employees. By working somewhat longer, but fewer days, employees spend less time and money commuting and enjoy an improved work–life balance. Their employers typically experience a drop in absenteeism, less overtime pay, and lower energy costs due to reduced use of heating, cooling and electricity. A one-year pilot study in Utah with 17,000 government employees working a four-day week reported a 13% reduction in energy use, an estimated 12,000 tonne reduction in GHGs, and savings for employees of up to $6 million in gas costs.

SOURCE
www.utah.gov/governor/news_media/
article.html?article=1724

GETTING STARTED

MEASURING GHGS

REDUCING GHGS

OFFSETTING GHGS

COMMUNICATING

MOVING FORWARD

Hudson's Bay Company
REDUCED ENERGY CONSUMPTION = LOWER OPERATING COSTS

Hudson's Bay Company (Hbc) is Canada's oldest company and one of its largest retailers, with nearly 600 stores, including the Zellers chain. Energy use from heating, cooling, and lighting its stores is a major operating cost. In 2000, Hbc committed to reducing the emission intensity of its energy use by 25% from 2000 levels by 2012.

To achieve its overall energy reduction target, Hbc worked with an energy consulting firm to focus on three main activities:

1. Retrofitting existing stores with T8 or more efficient lighting. By the end of 2006, 76% of Hbc locations had been retrofitted, reducing energy use by an average of 30% relative to the company's 2000 baseline.

2. Installing building automation systems in all stores. By the end of 2006, 74% of all Hbc stores had been retrofitted with systems to centrally control and monitor building functions such as lighting, temperature and humidity, in order to reduce energy consumption.

3. Installing energy efficient technology in all new stores and exceeding the Model National Energy Code for buildings by a minimum of 25%. For example, Hbc recently opened its greenest Zellers ever in Waterdown, Ontario, featuring a reflective white roof, energy recovery ventilators, high-efficiency HVAC, LED signage, wind turbines and solar panels.

Hbc estimates that each investment in energy efficiency will pay for itself within three years, taking into account incentives from governments and local utilities. So far, these measures are estimated to have reduced Hbc's energy costs by an estimated CDN $9.3 million annually, depending on energy prices and weather.

Hbc recognizes that it needs the support of its employees to maximize the potential of its energy reduction efforts. It has worked with local managers to make energy reduction an ongoing management issue, through education and incentives. Hbc also communicates regularly with its staff about its energy reduction programme in a number of ways, including an intranet system with discussion groups, in-store posters, regular newsletter articles and an annual social responsibility report that details Hbc's progress towards meeting its objectives.

In addition to its direct reductions in energy use, Hbc has committed to purchasing green power for five years to address the GHG emissions from remaining energy use.

FOR MORE INFORMATION
www.hbc.com

B. TRANSPORTATION

Transportation accounts for about 13% of global GHG emissions[11] and is therefore an important source of emissions for most businesses. Here are some examples of how businesses can reduce the climate impact of their transportation activities:

- **Implement an anti-idling policy** for all company vehicles.
- **Increase overall fleet efficiency**. Switch to smaller and more fuel-efficient vehicles capable of performing the tasks required ('right-sizing'); optimize delivery routes; ensure regular maintenance of vehicles.
- **Reduce emissions from commuter travel.** Promote carpooling; provide subsidies for public transit passes; locate offices near public transportation routes; provide secure parking for bicycles, and showers and changing rooms for cyclists; create telecommuting options for employees.
- **Reduce business travel**. Make sure trips are multi-purpose, and use video- and teleconferencing for meetings when possible. Some organizations are now hosting virtual conferences, where both speakers and participants attend online instead of travelling to a host city.
- **Choose the most climate-friendly transportation option** when possible: for example, use bike couriers and hybrid vehicle taxis, and trains instead of planes.
- **Locate new businesses and facilities at a minimum distance for suppliers, customers and employees.**

Companies working together

Food manufacturers **Nestlé** and **United Biscuits** are taking part in an unusual collaboration in the UK to save fuel and reduce emissions. United delivery trucks that were formerly empty on return trips are now delivering Nestlé products. So far the companies have together saved 280,000 kilometres, 95,000 litres of fuel and 250 tonnes of carbon dioxide each year. The only costs involved were a few hours of management time to organize the initiative.

SOURCE
www.climatechangecorp.com/content.
Asp?contentid=6209

Saving on business travel

Cisco, a global provider of networking solutions, has developed its Cisco TelePresence™ conferencing system to create a live, face-to-face communication experience without the need to travel. Reducing business travel means fewer related GHG emissions, but also saves money on travel-related costs, including air fare, hotels, and staff time. Traveling less can also improve employee productivity and quality of life.

SOURCE
www.cisco.com/telepresence

Stop idling, save money

Idling an average vehicle for ten minutes a day uses about 100 litres of gas per year. At $1.00 per litre, that represents $100 in potential savings per year just by turning off the engine.

SOURCES
www.crd.bc.ca/rte/idling.htm
www.idlefreebc.ca

GETTING STARTED MEASURING GHGs REDUCING GHGs OFFSETTING GHGs COMMUNICATING MOVING FORWARD

C. RENEWABLE ENERGY SOURCES

Businesses can reduce their GHG emissions by using renewable sources of energy. Some possibilities include:

- **Install micro wind turbines or solar energy panels** (hot water and/or photovoltaic). In some jurisdictions, it may even be possible to sell excess electricity back to the grid.
- **Use ground source heat exchange systems**, which provide low-cost heating and cooling. They are relatively easy to install in new construction and can also be retrofitted to existing buildings. These systems can transfer indoor heat into the earth during hot weather, cooling the building, and transfer heat from the earth into buildings during winter. They require only a small amount of electricity to function once installed.
- **Use heat recovery.** There are many ways that waste heat can be recovered, including from a building's own wastewater as it flows through a building's sewer pipes, or by using excess heat from a nearby facility.
- **Purchase renewable energy.** In some locations, it is now possible to purchase renewable energy from utility companies or from dedicated renewable energy providers. These companies add power to the local grid that is generated by renewable sources in an amount equal to that purchased by the customer. Renewable energy may be available as a bundled service in which the utility adds a green premium to its base electricity costs, or as an unbundled service where the business continues to purchase conventional electricity from the local utility but also makes a separate purchase of renewable energy certificates (RECs) from a green power provider (see *What's a REC?* box on this page).

What's a REC?

Renewable energy certificates, or RECs (sometimes also called renewable energy credits, green certificates, or green tags) are sold by utilities or green energy developers. A REC is a certificate issued when one megawatt-hour of electricity is generated and delivered to the grid from a qualifying renewable energy source, such as wind or solar. The REC can then be sold, separately from the electricity, to consumers who want to support green power, even if they do not actually receive electricity from the renewable energy source. It should be noted that RECs are not the same as carbon offsets.

SOURCE
www.offsetqualityinitiative.org/OQI%20 REC%20Brief%20Web.pdf

Working to be green

WorkCabin.ca, a website for green jobs, announced in 2007 that it was purchasing renewable energy certificates for all of its electricity needs. 'We know using the word "green" to describe what we do and who we are carries a major responsibility to show real action,' said WorkCabin founder Gregg McLachlan.

SOURCE
www.workcabin.ca

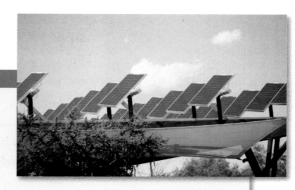

GETTING STARTED

MEASURING GHGs

REDUCING GHGs

OFFSETTING GHGs

COMMUNICATING

MOVING FORWARD

CASE STUDY

Johnson & Johnson

USING RENEWABLE ENERGY
TO REDUCE EMISSIONS

Johnson & Johnson is the world's largest medical devices and diagnostics company, and also has major pharmaceutical and consumer product divisions selling familiar brands such as BAND-AID® and JOHNSON'S® Baby Shampoo. In 2008, revenue from its 250 companies totalled $63.7 billion.

In 2003, Johnson & Johnson launched its 'Climate Friendly Energy Policy', which was designed to reduce both the company's climate impact and its operating costs. The policy established a goal to reduce the company's greenhouse gas emissions by 7% in absolute terms by 2010, using 1990 as a base year. It also laid out a comprehensive strategy for reducing emissions from energy use, including making $40 million in capital funding available each year for energy efficiency and greenhouse gas reduction projects across the company.

One of Johnson & Johnson's first priorities is to make engineering changes and equipment upgrades at company facilities to reduce energy consumption. The company also looks for opportunities to install cogeneration systems, where waste heat is recovered from on-site electrical generation to maximize overall efficiency.

Another way Johnson & Johnson reduces its emissions is through the use of renewable energy.

The company first identifies ways to increase its use of on-site renewable energy, including solar, wind and biomass. By the end of 2007, Johnson & Johnson had become the second largest corporate user of on-site solar power in the US, with more than 4.1 MW of solar photovoltaic capacity installed across ten company locations. To further increase its use of renewable energy, the company also makes direct purchases of low-impact hydro and wind from green power providers, and purchases renewable energy certificates.

By July 2009, 34% of Johnson & Johnson's total electricity needs in the US were being provided by green power, according to the US Environmental Protection Agency, making the company the eleventh largest user of green power in the US.

As a result of these initiatives and others, Johnson & Johnson achieved its greenhouse gas reduction target three years early: by 2007, it had reduced its emissions 12.7% from 1990 levels, despite the fact that company sales increased by 400% in the same period. It is also saving money: the investments in its own facilities and on-site renewable energy have generated on average a 17.2% rate of return.

FOR MORE INFORMATION
ww.jnj.com

Delivering reductions

A division of **FedEx**, FedEx Express is the world's largest express transportation company, delivering around 3.4 million packages daily in more than 220 countries. By improving its operational efficiency the company is reducing both its fuel use and greenhouse gas emissions. Its efforts have included optimizing its routes so that the most efficient-sized vehicle is used, and maximizing the density of its ground and air shipments so that fewer trucks and planes are needed to make the same number of deliveries. The company has already converted over a quarter of its fleet to smaller, more fuel-efficient vehicles, and the FedEx group of companies currently operates the largest fleet of commercial hybrid trucks in North America. FedEx Express estimates that in the US alone, since 2005, it has saved 45 million gallons of fuel and avoided 452,573 tonnes of CO_2 through these initiatives.

SOURCE
www.fedex.com

D. OPERATIONAL EFFICIENCY

Businesses and other organizations that improve operational efficiency can reduce their GHG emissions by minimizing the time and resources required to carry out their regular activities. Some examples include:

- **Optimize production.** It has been estimated, for example, that optimizing production processes and systems could mean energy savings of between 20 and 50% across industrial sectors in the United States.[12] Energy savings mean reductions in GHG emissions, as well as cost savings.
- **Improve efficiency related to logistics.** Businesses engaged in activities like deliveries to customers or the distribution of goods can improve efficiency and achieve GHG reductions at the same time – fewer kilometres travelled overall translates into lower GHG emissions, and lower fuel costs.

E. MATERIAL INPUTS

Another way to reduce GHG emissions is to reduce the materials that a business consumes in its operations, or change the materials so that they are less energy- or GHG-intensive. Some examples include:

- **Paper use.** The production of paper can be very energy-intensive, and businesses that use a lot of paper may find that this is a significant source of GHG emissions. It's easy to use less paper by relying on electronic documents where possible and by printing and photocopying double-sided, and only when necessary. In addition, switching to 100% post-consumer recycled paper not only saves trees and reduces waste directed to the landfill, but can also reduce the GHG emissions associated with paper by around 40%.[13]
- **Packaging.** Reducing the size or overall packaging of products can reduce both the GHG emissions from manufacturing the packaging and those from shipping the packaged goods.
- **Production inputs.** Substituting materials used in the production process can lower GHG emissions. For example, recycled pop bottles are now being used to manufacture fleece fabric for outdoor clothing.
- **Refrigerants.** HFCs (hydrofluorocarbons) are a chemical commonly used in refrigeration equipment. Because HFCs are a potent greenhouse gas,

even small leaks associated with regular use of refrigeration equipment can have a major climate impact. Switching to greener refrigerant technologies can thus greatly reduce GHG emissions.

- **Fuel-switching.** Using greener forms of fuel to power operations (such as biomass and biogas) can result in reductions in greenhouse gas emissions.

F. UPSTREAM GHG REDUCTIONS FROM SUPPLIERS AND CONTRACTORS

Businesses can also reduce upstream GHG emissions associated with materials and services supplied to it by other companies. Many larger companies have programmes to green their supply chains by providing support and incentives to suppliers, but smaller businesses can also make choices regarding suppliers and contractors that reduce the climate impact of their business. For example:

- **Choose low-carbon or carbon neutral suppliers.** Where transportation is an important source of emissions from suppliers, businesses can use suppliers that are closer geographically, that use less GHG-intensive modes of transport like rail, marine or hybrid vehicles, or that offer carbon neutral services. These preferences can be incorporated into the overall procurement policy for a business.

Making better paper choices

The **Environmental Defense** paper calculator allows users to quantify the environmental benefits of better paper choices. For example, using 100% post-consumer recycled paper instead of virgin paper means a reduction of more than one tonne of GHG emissions for every tonne of paper used.

Markets Initiative offers a database of 'eco-papers' with information on recycled content, FSC certification, and the bleaching process used as well as paper grade and weight, with contact details for paper merchants and printers.

SOURCES
www.edf.org/papercalculator
www.marketsinitiative.org/EPD/

Removing HFCs reduces GHG emissions

The Coca-Cola Company's (TCCC) largest emission source by far is its refrigeration equipment, with 9 million coolers and vending machines around the world. In 2006, the company switched to HFC-free insulation for this equipment, and is now working to remove the HFC refrigerants used to cool the equipment, and replace them with more climate-friendly CO_2. TCCC expects to have 100,000 of these HFC-free coolers and vending machines in operation by 2010. The company has also developed a proprietary Energy Management System that can reduce energy use by up to 35% by learning the usage patterns of the equipment. TCCC's new eKO equipment incorporates all three of these changes, and will reduce direct and indirect GHG emissions by more than 5 tonnes over each cooler's lifetime.

SOURCE
www.climate.thecoca-colacompany.com

- **Work with existing suppliers and contractors to find ways to reduce their emissions**, and ensure that price and other incentives reward low-carbon suppliers. This may also lead to cost savings for businesses in the form of less costly inputs or reduced distribution costs.

G. DOWNSTREAM REDUCTIONS

Businesses can also reduce their climate impact by using their influence to promote GHG emission reductions downstream from their operations, including after their products are sold to consumers. Here are some ideas:

- **Reduce the GHG emissions associated with the use of products.** For example, incorporate energy efficiency into product design (e.g. electronics, cars, etc.) so that they consume less energy over their lifetime.
- **Reduce the GHG emissions associated with the disposal of products.** Designing products so that they can be reused, disassembled, recycled, or composted can minimize the emissions from landfills, where rotting waste can produce methane, a very potent greenhouse gas. Similarly, reducing packaging not only lowers the emissions from manufacturing and shipping, but also the emissions created when the packaging is hauled by garbage trucks and landfilled.
- **Develop products and services that help customers reduce their own climate change impacts.** For example, some banks offer special loans that provide incentives for customers to perform energy retrofits of their home or business.

Additional ways for businesses to reduce their emissions can be found in the *Helpful resources* on the next page.

Drugstore cuts down on upstream emissions

Boots, a large drugstore chain, switched to using 30% post-consumer plastic in the bottles of selected products, leading to a 10% reduction in the overall carbon footprint of those products.

SOURCE
www.carbon-label.com/casestudies/Boots.pdf

Energy saving service

Ricoh Europe, a supplier of office equipment, sells and services a number of energy efficient photocopiers. But the company's service personnel found that its customers weren't letting their photocopiers switch to energy-saving standby mode when not in use – busy office staff found the warm-up time of one minute just too long to wait. In response, Ricoh worked with focus groups to determine an acceptable wait time and designed copiers that warm up in 10 seconds.

SOURCE
www.climatechangecorp.com/content.asp?ContentID=6131&ContTypeID=8

GETTING STARTED
MEASURING GHGs
REDUCING GHGs
OFFSETTING GHGs
COMMUNICATING
MOVING FORWARD

CASE STUDY

Resort Municipality of Whistler
MAKING REDUCTIONS AND SETTING
A CARBON NEUTRAL GOAL

The Resort Municipality of Whistler is a popular skiing and tourist destination in British Col-umbia, hosting the Vancouver 2010 Olympic and Paralympic Winter Games. In 2007, Whistler and other local governments in BC signed a Climate Action Charter,[14] committing to a goal of being carbon neutral by 2012, in conjunction with the province's own goal of carbon neutrality by 2010.

The carbon neutral goal fits well with Whistler's ongoing efforts to reduce its climate impact. In 1997, Whistler joined the Federation of Canadian Municipalities' Partners for Climate Protection (PCP) programme for local governments aiming to reduce their GHG emissions. Whistler's major corporate emission sources include its vehicle fleet (including transit), construction, municipal services and the operation of its facilities. Major community emission sources include passenger vehicles, and energy use by commercial and residential buildings.

Whistler's emissions reduction target is 12% below 2000 levels by 2012. Municipalities can reduce emissions directly, or indirectly through bylaws, building codes, and zoning. They can also work with local utilities, businesses and residents. Some of Whistler's measures include:

- A district energy system that uses heat captured from a wastewater treatment plant to meet space and hot water needs for a new residential neighbourhood.
- Switching the community's gas distribution system from propane to natural gas (which will reduce emissions by 15%).
- Industrial composting.
- Replacing 150,000 5-watt holiday lights with more efficient LED bulbs.
- Developing a municipal Green Building Policy, and WhistlerGreen building standard.
- Implementing a transportation demand management programme, improving the efficiency of its vehicle fleet, and creating facilities for cyclists.

Emissions from landfill gas have also been reduced significantly through a recently completed landfill 'cap and capture' project.

FOR MORE INFORMATION
www.whistler.ca

 ## Helpful resources
for reducing GHG emissions

A Business Guide to U.S. EPA Climate Partnership Programs
by U.S. Environmental Protection Agency
www.epa.gov/partners/Biz_guide_to_epa_climate_partnerships.pdf

The Carbon Trust
www.carbontrust.co.uk

Commuter challenge: getting commuters out of their cars

Commuter Challenge is a national programme that encourages Canadians to walk, cycle, take transit, carpool or tele-work instead of driving alone to work. It is based on a friendly competition between workplaces and communities to have the highest participation rates during the week of the event. Commuter Challenge supports workplaces as they encourage their employees to leave their cars at home. In 2009, 44,551 Canadians in over 159 communities and 1,636 workplaces participated in the Commuter Challenge. By registering online, they were able to see the greenhouse gas reductions they achieved.

SOURCE
www.commuterchallenge.ca

Cool Companies
by Center for Energy & Climate Solutions
www.cool-companies.org/homepage.cfm

What Can Be Done
by The Climate Group
www.theclimategroup.org/facts_and_actions/what_can_be_done

Guide to Purchasing Green Power: Renewable Electricity, Renewable Energy Certificates and On-Site Renewable Generation
by U.S. Environmental Protection Agency
www.epa.gov/grnpower/documents/purchasing_guide_for_web.pdf

Switching to Green: A Renewable Energy Guide for Office and Retail Companies *by World Resources Institute*
www.wri.org/publication/switching-green-renewable-energy-guide-office-and-retail-companies#

Green Electricity Marketplace *by Green Electricity Marketplace Ltd. (UK)*
www.greenelectricity.org

Green Power Partnership
by U.S. Environmental Protection Agency
www.epa.gov/grnpower

Database of State Incentives for Renewables and Efficiency (DSIRE)
www.dsireusa.org

U.S. Department of Energy – Energy Efficiency and Renewable Energy
www.eere.energy.gov

Natural Resources Canada – Office of Energy Efficiency
http://oee.nrcan.gc.ca

Business and Industry Resources
by Australian Government Department of Climate Change
www.climatechange.gov.au/resources/industry.html

Three Steps to Eco-Efficiency for Small and Medium-Sized Manufacturers
by Industry Canada
http://strategis.ic.gc.ca/epic/site/ee-ee.nsf/en/ef00012e.html

Making Your Data Centre Greener
by ZDNet
http://resources.zdnet.co.uk/articles/features/0,1000002000,39288042,00.htm

GETTING STARTED

MEASURING GHGs

REDUCING GHGs

OFFSETTING GHGs

COMMUNICATING

MOVING FORWARD

PART 4

Offsetting Greenhouse Gas Emissions and Going Carbon Neutral

Part 4 discusses how to create carbon neutral initiatives, including carbon neutral businesses, products, services, events and projects. It also explains how carbon offsets work and what to consider when purchasing them.

What is carbon neutral?

Carbon neutral (also known as climate neutral) refers to an organization or individual that has reduced the overall net climate impact of their operations to zero. This is usually a three-part process: measuring GHG emissions, reducing GHG emissions, and then offsetting remaining GHG emissions to become carbon neutral.

Measuring and reducing GHG emissions are activities that a business undertakes with respect to its own operations (see Parts 2 and 3). In order to go carbon neutral, a business will also need to purchase reductions, known as *carbon offsets*, from another source. For example, a business with total emissions of 100 tonnes for a one-year period (after its own direct reductions are taken into account) would need to purchase 100 tonnes of offsets to become carbon neutral. Each year, the process of measuring, reducing and offsetting is repeated.

'Carbon neutral' can also be a goal to be achieved within a set timeframe (e.g. becoming carbon neutral by 2010), which can allow a business more time to achieve its own direct reductions prior to the date set for becoming carbon neutral.

Who's going carbon neutral?
Businesses large and small are going carbon neutral. **News Corp., Interface, HSBC, Dell, Swiss Re, Eurostar, Marks & Spencer** and a host of others are managing and reducing their own GHG emissions and taking responsibility for their remaining climate impact by using carbon offsets.

Businesses are not the only ones taking action. The **United Nations** has set in motion an initiative to make all UN agencies and programmes climate neutral. The **World Bank** also has a carbon neutral program. **Norway, Costa Rica, Iceland** and **New Zealand** have all made country-wide carbon neutral commitments, as has the **Vatican**.

Going carbon neutral can also be a powerful way to communicate about climate impact and solutions to event spectators and fans. Bands like the **Rolling Stones, U2** and the **Dixie Chicks** have all offset the travel for their tours.

Olympic athletes are using carbon offsets to reduce the impact of their air travel, and both the **Super Bowl** and the **Australian Football League** have also gone carbon neutral.

SOURCE
www.davidsuzuki.org/Climate_Change/What_You_Can_Do/carbon_neutral.asp

DIAGRAM 6:
USING CARBON OFFSETS TO BECOME CARBON NEUTRAL

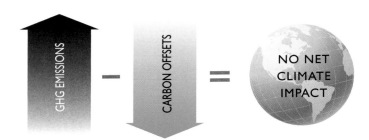

Making a decision to go carbon neutral

In theory, a business can become carbon neutral by purchasing offsets without making any effort to reduce its own emissions. In practice, however, this has been criticized as failing to take adequate responsibility for the business's own emissions, and relying on a simple purchase to address its climate impact.

To address these concerns and ensure the credibility of any carbon neutral initiative, this guide recommends that businesses look at their own climate impact and take steps to lessen it. Of course, as part of preparing a carbon neutral initiative, a business may purchase carbon offsets for some or all of its emissions at the same time as it is developing reduction opportunities. Over time, however, the guiding principle should be to reduce the number of offsets that need to be purchased, by finding more ways to reduce the business's own emissions.

Some of the main benefits and risks of going carbon neutral are discussed below. While carbon neutral initiatives will not suit every business, in the right circumstances both the business and the climate can benefit.

Benefits of carbon neutral initiatives

When carbon neutral initiatives include efforts to reduce the business's own GHG emissions, there are a number of potential benefits:

- **Going carbon neutral can be a powerful way to communicate** to employees, the public and other stakeholders that a business has made a commitment to manage its GHG emissions in a comprehensive manner, and show leadership on climate change.

- **A carbon neutral initiative can act as an interim measure to give a business time to make internal reductions**, while still taking responsibility for its climate impact.
- **A business can address the climate impact of all its GHG emissions**, including those it has not been able to reduce, and even those it does not directly control, such as those from suppliers.
- **Setting a carbon neutral goal and purchasing carbon offsets puts a price on the GHG emissions of a business** and this additional expense can help drive internal reductions.
- **High quality carbon offset projects create real reductions in global GHG emissions**, as well as bringing environmental and economic co-benefits to regions where the offset projects take place.
- **Businesses can offer consumers innovative carbon neutral products and services.**

Risks associated with carbon neutral initiatives

Along with the opportunities associated with carbon neutral initiatives, there are also certain risks:

- **There are currently no broadly accepted standards for carbon neutral claims,** so a business has nothing to fall back on except its own claim, although it may have its measurements and offsets verified by third parties.
- **Lower quality carbon offsets may have little or no climate benefit,** jeopardizing the integrity of the carbon neutral claim, and exposing the business that uses the offsets to reputational risk. See W*hat is a carbon offset?* (page 49) for more information.
- **Purchasing offsets may be costly.** Internal reductions by businesses are usually a one-time investment that delivers permanent benefits, whereas relying on carbon offsets means recurring costs at the end of each emissions inventory period. It is important to conduct an emissions inventory prior to any public commitment in order to fully understand the cost implications of the carbon neutral initiative.
- **Companies may be accused of 'buying their way out',** particularly when they simply purchase offsets and make claims of carbon neutrality without having also made efforts to reduce their own emissions.

Google carbon neutral
There are now hundreds of millions of **Google** users online, and the extensive computer infrastructure required to keep Google tools and services running uses a lot of electricity. In an effort to reduce its climate impact, Google has made a commitment to be carbon neutral every year, beginning in 2007. To accomplish this, Google is finding ways to reduce its own energy use by improving the energy efficiency of its data centres and offices. It has also set a goal of building one gigawatt of new renewable energy capacity (enough electricity to power a city the size of San Francisco), focusing on thermal solar energy. Finally, to take full responsibility for its remaining footprint, Google is investing in offset projects around the world that reduce GHG emissions.

SOURCE
www.google.ca/intl/en/corporate/green/footprint.html

GETTING STARTED

MEASURING GHGs

REDUCING GHGs

OFFSETTING GHGs

COMMUNICATING

MOVING FORWARD

Developing an effective carbon neutral initiative

With the right approach a business can maximize the benefits from a carbon neutral initiative and minimize the risks. Here are some key factors to consider:

- **Broad emissions scope.** Because the term 'carbon neutral' implies zero net climate impact, it is essential that the emissions scope be as broad as possible, and include all scope 1 and 2 emissions, as well as all major, relevant, scope 3 emissions. To ensure that all major emission sources are

CASE STUDY

Vancity

FIRST NORTH AMERICAN-BASED CARBON NEUTRAL FINANCIAL INSTITUTION

Vancity is Canada's largest credit union, with CDN $14.1 billion in assets and almost 400,000 members. Vancity is known as an innovator in the financial sector, and has also implemented many social and environmental programmes in the course of its business, such as its Clean Air Auto Loan, and low-interest loans for energy efficiency retrofits for homes and businesses.

In 2005, Vancity made a commitment to make its entire operations carbon neutral by 2010. However, in 2008, Vancity announced it had already achieved its carbon neutral goal – the first North American-based financial institution to do so.

The Vancity carbon neutral programme is an ongoing initiative with three elements:

1. Measuring. Using the GHG Protocol as guidance for determining the boundaries for its inventory, Vancity included four major sources of emissions: energy use from buildings, business travel by staff, staff commuting, and paper use. In 2007, these sources totalled 6,010 tonnes of CO_2e.

2. Reducing. Vancity has been working to reduce its GHG emissions since 1990, and estimates its

carbon footprint is 50% smaller per employee than other Canadian financial institutions. It has made its buildings more energy efficient, and has saved $2 million in energy costs as a result. It has reduced paper use by 30% since 1997, and uses primarily 100% post-consumer recycled paper. As well, Vancity provides a number of incentives to employees to get them out of their cars; as a result, a majority of Vancity employees now commute to work by public transit, walking, cycling or ride-sharing.

3. Offsetting. Vancity worked with external environmental experts and developed its own guidelines to ensure that all offsets it purchases are of high quality, and it uses only offsets from renewable energy and energy efficiency projects. In 2007, Vancity launched a carbon offset granting programme which aims to support local offset projects.

As a way to broaden the impact of its carbon neutral programme Vancity offers financial incentives and resources to its individual and business customers, to assist them in reducing their GHG emissions.

FOR MORE INFORMATION
www.vancity.com

included, businesses need to be aware of their overall climate impact by conducting an emissions inventory.

- **Transparency**. Businesses need to be as transparent as possible about their carbon neutral claims to ensure credibility with customers and other stakeholders. Information should be easily accessible and include which emissions sources are being offset, the quantity, any major emissions sources that were not included (and why), the relevant time period, methodologies used to calculate the emissions and how the emissions were verified, and details about the carbon offsets being used.

- **Reducing emissions**. The carbon neutral initiative should be part of a wider greenhouse gas management programme that includes reductions in the business's own GHG emissions.

- **High quality carbon offsets**. Businesses should only use high-quality carbon offsets, like those certified to the Gold Standard, and should be prepared to carry out due diligence with respect to any offsets they purchase, particularly in the voluntary carbon market. More guidance on offsets is provided in *What is a carbon offset?* (page 49).

- **Educational component**. The carbon neutral initiative should be designed to educate employees, customers and/or suppliers about the climate impact of the emissions associated with the business's products and services, as well as opportunities for reductions.

- **Credible claims**. Caution should be exercised in making blanket claims of carbon neutrality (for example, 'our business is now carbon neutral'), particularly when a business hasn't examined its entire carbon footprint or made efforts to offset major scope 3 emissions. Similarly, a business that measures and offsets its emission sources, but nonetheless has a business model that is GHG-intensive or creates products whose use generates significant emissions (for example, vehicles), should carefully consider the risk of negative publicity before making a claim of carbon neutrality.

- **Third-party review**. A business making public claims about its carbon neutral initiatives should obtain independent assurance that it has accurately measured its emissions, and accounted for all relevant emissions sources. In general, this would involve verification or review of the calculations by a reputable organization experienced in greenhouse gas auditing.

Climate neutral from 'cow to cone'

In April 2007, **Ben & Jerry's** went climate neutral from cow to cone on all ice-cream flavours it produces in Europe. In examining its carbon 'hoofprint', the company included emissions from dairy farming, ingredient sourcing, manufacturing, packaging, transport, and freezer equipment. Reduction initiatives are also underway in all parts of its supply chain. For emissions that cannot currently be avoided, Ben & Jerry's is using Gold Standard carbon offsets.

SOURCE
www.cleanair-coolplanet.org/documents/zero.pdf

GETTING STARTED

MEASURING GHGs

REDUCING GHGs

OFFSETTING GHGs

COMMUNICATING

MOVING FORWARD

• **Standards**. To date there is no generally accepted standard for carbon neutral initiatives, although there are several standards available for carbon offsets. It is likely that one or more carbon neutral standards or protocols will emerge as leaders over time, and businesses should be prepared to consider having their carbon neutral initiatives evaluated according to one of these standards.

Different types of carbon neutral initiatives

There are a number of different forms that a carbon neutral initiative can take, including making a product or a service – or the entire business – carbon neutral. Some companies opt for a combination of these; for example, making their entire business, as well as key products, carbon neutral.

In each case the GHG emissions that are being measured and offset need to be carefully defined. A carbon neutral claim by a business implies a comprehensive effort to manage its GHG emissions. A business should broadly assess all of the significant emission sources associated with its claim, even though some of the emissions will likely be outside its control or influence.

For more guidance on scope 3 emissions, see *Which emission sources should be included in the inventory?* in Part 2 (page 12).

The table below describes five common types of carbon neutral initiatives. It provides a possible scope of emissions for each type of initiative, and outlines some further considerations.[15]

Carbon neutral events

For information about how to plan carbon neutral conferences and other events, see the David Suzuki Foundation web page:

www.davidsuzuki.org/Climate_Change/What_You_Can_Do/carbon_neutral_events.asp

---

Chef serves up a carbon neutral TV series

Jamie Oliver, TV chef and cookbook author, has put climate solutions on his menu. *Jamie's American Roadtrip* may be the world's first carbon neutral TV cooking and travel series. Total emissions for air travel, vehicle use and hotel stays were calculated to be 242.6 tonnes of CO_2. To compensate, Oliver's company invested in three offset projects: energy efficient stoves in Cambodia, wind power in China, and solar power in India. 'My company is serious about helping to prevent the dangers of climate change,' said Oliver, 'and we recognize that with everything we do – but especially with TV programmes like this one – we create a carbon footprint.' Oliver and his team will be looking for ways to reduce emissions from future series.

SOURCES
www.jamieoliver.com/news/jamie-s-america-roadtrip-offsets-co2-emi
http://news.uk.msn.com/uk/article.aspx?cp-documentid=149453839

TABLE 4: DIFFERENT TYPES OF CARBON NEUTRAL INITIATIVES

**CARBON
NEUTRAL
BUSINESS**

**CARBON
NEUTRAL
PRODUCT**

Examples of carbon neutral businesses
A carbon neutral retail store, a carbon neutral accounting firm.

Factors to consider
Going carbon neutral can be an iterative process whose scope can be expanded each year. A business may begin with internal reductions and offsetting one business unit or activity, such as air travel by executives, and work towards a full-scale carbon neutral initiative. Viewing carbon neutrality as an ongoing process can also help drive internal emission reductions over the long term.

This approach is generally more suitable for service sector businesses with less GHG-intensive activities, but there are also opportunities for innovative manufacturers and other businesses.

Possible scope of GHG emissions
All direct emissions (scope 1), use of purchased electricity, heat and steam (scope 2), and all relevant indirect emissions (scope 3).

Examples of carbon neutral products
Carbon neutral book, carbon neutral ice cream.

Factors to consider
A customer looking at a product labelled 'carbon neutral' likely has a reasonable expectation that the product itself, as a whole, has a *net zero* or very minimal impact on the climate. This means that the minimum scope of GHG emissions that should be included is likely broader than that for a carbon neutral business, where the focus is more on the activities of the business itself.

To calculate a product's emissions, a business will often start with its own emissions and then work with suppliers to trace additional emissions up the supply chain. Downstream emissions, such as those from delivering the product to customers (or even customer travel to stores), and from the use and disposal of the product should also be considered.

Creating a carbon neutral product can be a marketing advantage for a business, but it can also be an opportunity to reduce GHG emissions not directly controlled by the business. For example, in the process of measuring (and learning about) the emissions from the materials used to create the product, a business can engage its suppliers and find ways to make reductions.

It's important to note that it may not be feasible to measure 100% of the life-cycle emissions associated with any given product. Businesses should be prepared to identify and measure all significant life-cycle emissions. With respect to very small emission sources, time and money spent to measure them might be better used to reduce emissions from more significant emission sources associated with the product.

Possible scope of GHG emissions
Emissions associated with the life cycle of the product, including direct emissions from the business (scope 1 and scope 2), and all significant indirect emissions (scope 3) such as supply chain emissions, emissions from delivery, use and disposal of the product.

GETTING STARTED

MEASURING GHGs

REDUCING GHGs

OFFSETTING GHGs

COMMUNICATING

MOVING FORWARD

TABLE 4: DIFFERENT TYPES OF CARBON NEUTRAL INITIATIVES *continued*

CARBON NEUTRAL USE OF PRODUCT

Examples of carbon neutral use of products
Purchasing an automobile for which the manufacturer or dealer has offset fuel use for one year, or a computer that comes with the option of purchasing offsets for its use.

Factors to consider
Customers can be offered the option of purchasing offsets for a fixed period of use of the product, or the cost of those offsets can be automatically included in the purchase price.

While these types of initiatives can help educate customers about the GHG emissions associated with using various products, they are sometimes criticized for placing little onus on the manufacturer to take responsibility for their own emissions. To give this kind of initiative more credibility, manufacturers should consider reducing and offsetting the emissions associated with their own operations, including product manufacturing.

Possible scope of GHG emissions
Emissions associated with the use of a product by the consumer (scope 3), and possibly emissions associated with manufacture (scope 1, 2 and 3) and disposal (scope 3) of the product.

CARBON NEUTRAL SERVICE

Examples of carbon neutral services
A carbon neutral courier, a carbon neutral web hosting service, a carbon neutral flight.

Factors to consider
These initiatives can help to educate customers about the climate impact of the service if the business undertakes effective communications around its initiative.

There are two broad categories of carbon neutral services. The first is where a business gives its customers the option to purchase offsets themselves. These types of voluntary programmes have been very popular with airlines: at the time of purchasing a ticket for a flight, a customer is invited to purchase offsets selected by the airline. However, participation rates by customers can be low, especially if the initiative is not marketed very well.

It is also possible for a business to offset all emissions associated with its service, without relying on customers to do so. While this guarantees the offsetting will occur, it is still important to inform customers about the initiative (e.g. through marketing, or on the invoice) to take advantage of the opportunity to engage customers with climate change solutions.

Possible scope of GHG emissions
Emissions associated with providing the service or a clearly defined part of the service to the customer (could be scope 1, 2 and/or 3).

GETTING STARTED

MEASURING GHGs

REDUCING GHGs

OFFSETTING GHGs

COMMUNICATING

MOVING FORWARD

TABLE 4: DIFFERENT TYPES OF CARBON NEUTRAL INITIATIVES *continued*

CARBON NEUTRAL EVENTS AND PROJECTS

Examples of carbon neutral events and projects

A carbon neutral conference, a carbon neutral sporting event, a carbon neutral concert, a carbon neutral film.

Factors to consider

Making an event (e.g. a conference) or a project (e.g. a film) carbon neutral can give businesses an opportunity to gain experience with GHG management on a smaller scale before expanding it to broader operations. As well, both events and projects can actively engage large numbers of people with the carbon neutral initiative. For example, event advertising and information on tickets can be used to inform event attendees and sponsors about the overall initiative, and to promote related programmes such as public transit and waste reduction efforts at the event. Projects can also provide convenient opportunities to involve and educate employees in GHG management, and if they have a public component (like films, for example) they can also reach broader audiences.

Because projects and events usually have tight schedules, it is helpful to get a full commitment from senior management to the carbon neutral initiative when the planning starts, including adequate resources with respect to staff and funding. Ways to reduce emissions should also be incorporated into the initial planning stages so that there is time to achieve the reductions.

Possible scope of GHG emissions

EVENTS: Energy consumed by event venues (scope 1 and 2); travel to the host city by participants; local road transportation; and energy used during hotel stays by participants (scope 3). Smaller emission sources include transportation of goods for the event, event organizer travel during planning and preparation, energy consumed by the organizing office, paper use and waste generation.

PROJECTS: Emission sources to be included are usually similar to those from an ongoing business operation (i.e. all scope 1 and scope 2 emissions, and relevant scope 3 emissions), except that they will be for a time period defined by the start and finish of the project.

What is a carbon offset?

Because all carbon neutral initiatives rely on the use of carbon offsets, it is important to have an understanding of what an offset is, how offsets are produced, and how to ensure any offsets purchased are of high quality.

A carbon offset is simply a reduction in GHG emissions created by one party that can be purchased and used to balance the emissions of another party. Carbon offsets are a market-based solution to climate change based on two principles:

1. The location where GHG reductions occur is not important with respect to the climate impact, because greenhouse gases are quickly diffused around the globe once they enter the atmosphere; and

News Corporation
BECOMING CARBON NEUTRAL AND ENGAGING AUDIENCES WITH CLIMATE SOLUTIONS

News Corporation is a global media company founded by Rupert Murdoch, with properties in film, television, cable, magazines, newspapers, publishing, websites and more.

In 2007, News Corporation announced a plan to address its energy use and climate impact. Three goals were identified: first, to lower its energy use and switch to renewable sources (enough to reduce greenhouse gas emissions 10% by 2012, using 2006 as a base year); second, to become carbon neutral by 2010; and third, to engage its business partners, 53,000 employees, and millions of readers, viewers and web users around the world on the issue of climate change.

News Corp. has measured its carbon footprint annually since 2006. In 2008, its footprint was 642,237 tonnes of CO_2e, including electricity and fuel use, and business air travel. It has also analysed the supply chain footprint of key products, including DVDs, newspapers, and television and film productions. For example, Twentieth Century Fox Home Entertainment partnered with Wal-Mart to measure the emissions from the production, manufacture and distribution of DVDs, which came to about 340 grams of CO_2e per DVD.

Reducing emissions has been the number one priority. Some of News Corp.'s reduction efforts include: making its offices, facilities, and data centres more energy efficient; building a LEED-certified FOX Studios facility; using more fuel-efficient vehicles; and purchasing renewable energy.

News Corp. has built its carbon neutral programme gradually, to allow time to make its own reductions before purchasing carbon offsets. Working in stages has also allowed the company and its subsidiaries and affiliates to learn as they go. In 2004, the Twentieth Century Fox production, *The Day After Tomorrow*, became the first carbon neutral film. In 2006, British Sky Broadcasting became carbon neutral. In 2009, the entire seventh season of the FOX-produced hit TV show *24* became carbon neutral. FOX also produced the first-ever carbon neutral Emmy Awards show, as well as a carbon neutral DVD, *Futurama*. News Corp. began purchasing carbon offsets (including Gold Standard-certified offsets) on an annual basis in 2007, and will increase purchases to meet its 2010 carbon neutral commitment.

Moving beyond its own carbon neutral efforts and engaging audiences is an important goal for News Corp. Says chairman and CEO Rupert Murdoch, 'Our global reach gives us an unprecedented opportunity to inspire action from all corners of the world.' Initiatives to date have included: running campaigns like 'Green it. Mean it.' to educate audiences on ways to save energy; delivering millions of free energy efficient light bulbs to TV and newspaper subscribers; and creating the *FOX Green Guide* (FOXGreenGuide.com), an online resource for best practices in low-carbon and environmentally friendly film, television and event production.

FOR MORE INFORMATION
www.newscorp.com

GETTING STARTED
MEASURING GHGs
REDUCING GHGs
OFFSETTING GHGs
COMMUNICATING
MOVING FORWARD

Offset prices

Like offset quality, offset prices vary considerably, depending on factors such as the project type, the country in which the project is developed, the volume of offsets generated, whether the project is verified by third parties, and supply and demand.

In 2007, offsets on the voluntary market sold for a volume-weighted average price of $6.10 per tonne (this was an increase from $4.10 in 2006). While price is no guarantee of quality, the price of offsets from high quality offset projects tends to be higher than the average price. Gold Standard offsets, for example, could typically be purchased in 2007 for a price ranging from $15–$34+ per tonne.

SOURCES
Ecosystem Marketplace and New Carbon Finance, *Forging a Frontier: State of the Voluntary Carbon Markets 2008*, p. 8. www.theclimategroup.org/assets/resources/Top_10_-_Carbon_Offsetting.pdf

2. The costs involved in reducing greenhouse gas emissions vary across economic sectors and activities around the world. At present, every business will encounter a threshold beyond which it is too expensive or difficult to make further reductions in its own emissions. However, by purchasing carbon offsets a business can create more reductions in GHG emissions than would otherwise be possible, through a more cost-effective offset project somewhere else.

Many activities have the potential to generate carbon offsets. Renewable energy sources such as wind farms or installations of solar panels can create carbon offsets by displacing the use of conventional energy sources like coal. Energy efficiency projects can also create offsets by conserving energy and reducing the need to burn fossil fuels. Other potential offset projects include storing or capturing greenhouse gases, such as planting trees or capturing methane from agricultural processes or landfills, or destroying greenhouse gases resulting from certain manufacturing processes.

Regardless of their source, carbon offsets are generally quantified and sold to purchasers in metric tonnes of CO_2e. Offsets are available for purchase from vendors, brokers, or directly from project developers. Businesses can make offset purchases on an annual basis, or enter into a long-term purchasing agreement.

Because it is difficult to become carbon neutral without the use of carbon offsets, they will play an important role in any carbon neutral initiative. However, it should be noted that businesses can use offsets for a variety of purposes other than just carbon neutral initiatives: for example, to address the climate impact of a business's GHG emissions without making a carbon neutral claim, or to meet regulatory obligations.

The voluntary carbon market

At present there are a number of regulatory markets for carbon offsets, like the international market created under the Kyoto Protocol, or the EU Emissions Trading System. In addition to these regulatory markets there is also a much smaller but growing voluntary carbon market. As its name suggests, this market includes all offset purchases made by individuals, businesses and organizations who choose to take responsibility for their climate impact, even though they aren't required to do so by government regulation.

The offsets sold on the voluntary market come from a variety of projects around the world. The buyer side of the voluntary market is dominated by businesses, which in 2008 accounted for at least 66% of purchases.

It has been estimated that 123 million tonnes of CO_2e were transacted in the voluntary market in 2008 – almost double the amount in 2007.

SOURCE
http://ecosystemmarketplace.com/documents/cms_documents/StateOfTheVoluntaryCarbonMarkets_2009.pdf

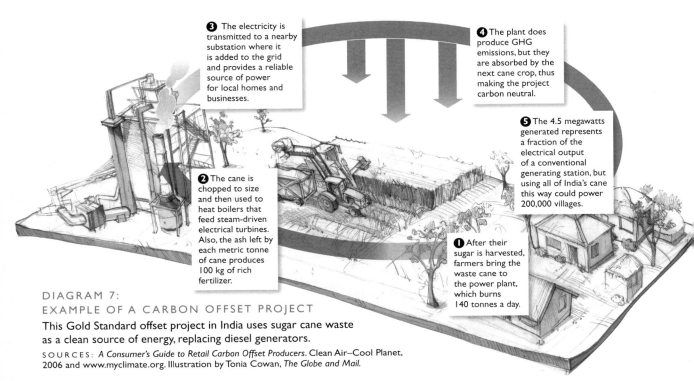

❸ The electricity is transmitted to a nearby substation where it is added to the grid and provides a reliable source of power for local homes and businesses.

❹ The plant does produce GHG emissions, but they are absorbed by the next cane crop, thus making the project carbon neutral.

❺ The 4.5 megawatts generated represents a fraction of the electrical output of a conventional generating station, but using all of India's cane this way could power 200,000 villages.

❷ The cane is chopped to size and then used to heat boilers that feed steam-driven electrical turbines. Also, the ash left by each metric tonne of cane produces 100 kg of rich fertilizer.

❶ After their sugar is harvested, farmers bring the waste cane to the power plant, which burns 140 tonnes a day.

DIAGRAM 7:
EXAMPLE OF A CARBON OFFSET PROJECT

This Gold Standard offset project in India uses sugar cane waste as a clean source of energy, replacing diesel generators.

SOURCES: *A Consumer's Guide to Retail Carbon Offset Producers.* Clean Air–Cool Planet, 2006 and www.myclimate.org. Illustration by Tonia Cowan, *The Globe and Mail.*

How do voluntary offset purchases help solve the problem of global warming?

To address global warming we will need to reduce GHG emissions in every sector of the economy. This will require concerted international cooperation and strong leadership from all levels of government. But it will also be important for businesses, other organizations and individuals to take responsibility for their own climate impact.

High quality offsets can create real and lasting reductions in GHG emissions, and thus can be an effective bridging strategy for businesses and other organizations as they work to reduce their climate impact. When voluntarily purchased, carbon offsets can also help address the gaps in existing regulations to limit greenhouse gases, and allow businesses to demonstrate leadership on global warming.

HIGH QUALITY CARBON OFFSETS

The voluntary carbon market offers prospective purchasers a wide range of offset project types, prices and quality. Businesses must be prepared to exercise due diligence, and have a good understanding of the quality issues with respect to carbon offsets, particularly if they choose to purchase offsets that are not verified to a recognized independent standard. Quality is critical to ensure that the investment in offsets accomplishes its primary purpose: to mitigate climate change by achieving new and lasting reductions in greenhouse gas emissions.

Businesses that use lesser quality offsets put their reputations at risk, and a carbon neutral initiative or offset purchase designed to create positive public relations can have the opposite effect, resulting in disillusioned customers, loss of investor confidence, and heightened scrutiny of the business's activities in general. Businesses should also keep in mind the growing knowledge and expectations of customers and shareholders with respect to climate change issues.

Quality issues around carbon offsets are relatively complex, and most businesses will likely not have the expertise or resources to adequately assess the quality of a given offset project. However, all organizations that use carbon offsets should at the very least be aware of basic quality considerations. See the *Helpful resources* beginning on page 55 for more information.

GETTING STARTED

MEASURING GHGs

REDUCING GHGs

OFFSETTING GHGs

COMMUNICATING

MOVING FORWARD

CASE STUDY

World Cup Soccer

USING GOLD STANDARD OFFSETS TO RAISE THE BAR FOR CARBON NEUTRAL EVENTS

The Fédération Internationale de Football Association (FIFA) World Cup is one of the largest sporting events in the world. In 2006, Germany hosted the first climate neutral World Cup. Organizers took into account major GHG emission sources associated with the event, including significant ones that are often ignored, such as venue construction and air travel by spectators.

Reduction efforts were focused on energy use and transportation. Organizers reduced energy use at all 12 World Cup stadiums (by an average of 13%) through efficient lighting management, heat recovery and other measures. Photovoltaic panels were installed at several stadiums, and organizers purchased additional green electricity for use by stadiums, hospitality facilities and the International Broadcasting Centre.

Organizers also promoted public transit for spectators travelling to the stadiums by ensuring frequent service, restricted parking at stadiums, and free travel for ticket holders on match days.

To address remaining emissions, and in keeping with the global spirit of the World Cup, the organizers purchased approximately 100,000 tonnes of Gold Standard offsets from projects in developing countries. As the organizers pointed out, 'the high standards of the projects are the most important factor in voluntary climate compensation, and they represent both a model and a challenge for future large sporting events'.

FOR MORE INFORMATION
www.oeko.de/oekodoc/292/2006-011-en.pdf

How to decide which carbon offsets to purchase

In light of the quality issues with carbon offsets, businesses should choose their offsets carefully. Following are some suggestions:

- **Consult expert reviews and analyses of carbon offsets and the voluntary carbon market.** See *Helpful resources for offsetting and going carbon neutral* (page 55) for a list of expert reports, including comparisons of offset vendors.
- **Seek advice from a reputable environmental organization or GHG consultant** that can provide information about carbon offsets. Most offset vendors will also offer their views on offsets but their advice may or may not be objective, and they might not be fully informed about quality issues. However, it can be helpful to talk to a number of different offset providers and ask them how their offset projects address the quality issues discussed in

Climate neutral parcels
DHL, a global logistics firm, offers climate neutral parcels through Deutsche Post. Customers pay a postal fee that includes climate neutral shipping. DHL calculates the CO_2 emissions created by the shipment of each parcel and offsets them through domestic and international offset projects.

SOURCE
www.dhl.com/publish/g0/en/press/
release/2008/080408.high.html

Particular quality issues with carbon offsets from tree-planting projects

While planting trees can have many benefits, offset projects based on planting trees face significant challenges. First, trees take many years to reach maturity, meaning that they do not deliver actual reductions in CO_2 from the atmosphere for many years – possibly decades – after they are planted. Secondly, while trees can be long-lived, they inherently lack permanence. Large amounts of carbon stored in trees can quickly be re-released as a result of forest fires, logging or disease. As well, some forests around the world are already turning into net sources of emissions because of the effects of global warming, creating further uncertainty about the climate benefit of this type of offset.

Organizations that purchase offsets from tree planting must be prepared to accept and manage an additional level of risk and uncertainty related to their carbon neutral initiatives. The UK band **Coldplay**, for example, offset the GHG emissions from its second album by having 10,000 mango trees planted in India. British media later reported that a large number of the trees had died because of a drought, creating negative public relations issues for the band.

SOURCES
www.davidsuzuki.org/Climate_Change/What_You_Can_Do/trees3.asp
www.telegraph.co.uk/news/worldnews/asia/india/1517031/How-Coldplay's-green-hopes-died-in-the-arid-soil-of-India.html

Can any business create and sell offsets?

A common question from businesses that find ways to reduce their GHG emissions is whether those reductions can be sold as offsets. It should be understood that developing offset projects is a technical undertaking requiring investment and specialized expertise. For example, offsets should be verified by a qualified auditor according to a recognized methodology. As well, a business cannot sell offsets from its own reductions if it counts the same reductions in its inventory, because it would be double-counting the same reductions. Nor would reductions from activities that are already underway or planned to occur be eligible because they would not meet the test of additionality. For more guidance, see the *Helpful resources* beginning on page 55.

this guide, as well as how much of their revenue goes directly to the project and how much covers their administrative costs.

- **Purchase offsets registered to a reputable standard.** This takes the guesswork out of selecting carbon offsets, and lends credibility to the carbon neutral initiative of a business. For example, standards such as the Gold Standard (see box on next page) help to ensure quality criteria are addressed, and that the offset projects have been audited by an accredited third-party organization. At present there are a number of standards emerging in the voluntary market, including some government standards. For more information, see *Helpful resources for offsetting and going carbon neutral* (page 55).

- **Review your offset purchasing policy on a regular basis.** A company's approach to greenhouse gas management will no doubt change over time, and the carbon market is also changing rapidly. Regular review will ensure that offset purchases are still in line with best practice.

The Gold Standard for carbon offsets

The Gold Standard is designed to ensure that carbon offsets provide real reductions in greenhouse gas emissions and that the carbon offset projects promote sustainable development objectives in the host communities. Gold Standard offset projects must be either renewable energy or energy efficiency projects, and verified by UN-accredited auditors according to recognized methodologies.

The Gold Standard was developed by **WWF**, SSN and Helio International and is now an independent non-profit organization based in Switzerland supported by a broad group of stakeholders, including more than 60 non-governmental organizations worldwide. It is consistently ranked among the highest standards for carbon offsets in the world by analysts. Gold Standard offsets have been used by high-profile businesses and organizations around the world including **HSBC**, **Ben & Jerry's**, **KLM**, **FIFA World Cup 2006**, **National Hockey League Players' Association**, and **UPS**.

SOURCE
www.cdmgoldstandard.org

Helpful resources for offsetting and going carbon neutral

CARBON NEUTRAL INITIATIVES

Three Stage Approach to Developing a Robust Offsetting Strategy
by Carbon Trust
www.carbontrust.co.uk/publications/publicationdetail?productid=CTC621

Getting to Zero: Defining Corporate Climate Neutrality
by Clean Air–Cool Planet and Forum for the Future
www.cleanair-coolplanet.org/documents/zero.pdf

Kick the Habit: A UN Guide to Climate Neutrality
by United Nations Environment Programme
www.unep.org/publications/ebooks/kick-the-habit/Pdfs.aspx

Neutral & Beyond, A Review of Carbon Neutrality and Offsets
by M. Lynch et al, Green Capital, an Initiative of Total Environment Centre
www.greencapital.org.au/index.php?option=com_docman&task=doc_download&gid=86

Resources for Going Carbon Neutral
by David Suzuki Foundation
www.davidsuzuki.org/Climate_Change/What_You_Can_Do/carbon_neutral.asp

Offsetting air travel

Many airlines are offering customers carbon offsets for their flights. For example, customers flying on **Virgin Atlantic Airways** have the option to purchase carbon offsets at the time of booking or even when they purchase duty free goods on board the plane. Virgin has calculated the average emissions for all of its flights, and secured a supply of Gold Standard offsets for its customers. In Vancouver, **Harbour Air Seaplanes**, a small airline offering regional flights, purchases carbon offsets to mitigate the climate impact of all scheduled service, charter and tour operations. The price of the offsets is automatically included in the price of all flights.

SOURCES
www.virgin-atlantic.com
www.harbour-air.com

GETTING STARTED

MEASURING GHGs

REDUCING GHGs

OFFSETTING GHGs

COMMUNICATING

MOVING FORWARD

First Chinese carbon neutral company

In 2009 **Tianping Auto Insurance** became the first Chinese company to purchase domestic offsets and become carbon neutral. The Shanghai-based company purchased 8,025 tonnes of carbon offsets for $40,627 from the China Beijing Environment Exchange, an emissions trading platform. To date China has been the leading supplier of carbon credits to the global market, but buyers have been from other countries. This may mark the beginning of domestic carbon trading in China.

SOURCE
http://en.cop15.dk/news/view+news?newsid=1842

SUPPLY CHAIN EMISSIONS AND CARBON LABELLING

Carbon Footprints in the Supply Chain: The Next Step for Businesses
by Carbon Trust
www.carbontrust.co.uk/publications/publicationdetail?productid=CTC616

The Carbon Reduction Label
www.carbon-label.com

CarbonCounted
www.carboncounted.com

CARBON OFFSETS

Top 10 Tips for Purchasing Carbon Offsets
by The Climate Group
www.theclimategroup.org/assets/resources/Top_10_-_Carbon_Offsetting.pdf

Purchasing Carbon Offsets: A Guide for Canadian Consumers, Businesses and Organizations
by David Suzuki Foundation and the Pembina Institute
www.davidsuzuki.org/Publications/offset_vendors.asp

Evaluations and Recommendations of Voluntary Offset Companies
by Tufts University Climate Initiative
www.tufts.edu/tie/carbonoffsets/TCI-offset-handout.htm

A Consumer's Guide to Retail Carbon Offset Providers
by Clean Air–Cool Planet
www.cleanair-coolplanet.org/ConsumersGuidetoCarbonOffsets.pdf

Carbon Offset Watch 2008 Assessment Report
by C. Riedy and A. Atherton, The Institute for Sustainable Futures, University of Technology
www.carbonoffsetwatch.org.au

Carbon Offset Research & Education
by Stockholm Environment Institute
www.co2offsetresearch.org/consumer/index.html

Offset Quality Initiative
www.offsetqualityinitiative.org

PART 5

Communicating Effectively: Supporting the GHG Management Programme

Part 5 discusses the importance of communications to the success of the GHG management programme, and highlights the key elements of an effective communication plan. It also looks at how to tailor messages to different audiences, and includes advice on how to avoid greenwashing.

When a business implements a GHG management programme, it is important that any related communications help to maximize the programme's benefit for the organization, and also its effectiveness as an environmental initiative. More than one business has implemented an innovative GHG management programme, only to see it languish because of a perceived lack of stakeholder support – when in fact stakeholders were not even aware of the initiative because of poor communications.

Some of the specific benefits of effective communications include:

- **Motivating and engaging employees** to participate in achieving the goals of the programme.
- **Informing stakeholders** of the company's progress in meeting its targets.
- **Helping create positive brand visibility** among investors and customers.
- **Educating and inspiring customers and the broader public** to take action on climate change.
- **Gaining positive media coverage** for the organization.
- **Inspiring other businesses and organizations** to take action.
- **Addressing regulatory reporting requirements**, if applicable.

Informing customers about climate impact

To help its customers understand the climate impact of its shoes, **Timberland** ranks them on a climate-impact scale of zero to ten, and this information appears on tags and leaflets inside shoe boxes. This information is based on Timberland's emissions inventory, which showed that 79% of the life-cycle emissions associated with its footwear result from the livestock used in the production of leather.

In the UK, the **Carbon Trust** has developed a Carbon Reduction Label. Participating companies measure the embodied GHG emissions from products and services across their life cycles, commit to reducing those emissions, and put labels on their products indicating the emissions per unit of the product or service.

SOURCES
www.cleanair-coolplanet.org/documents/zero.pdf
www.boston.com/business/articles/2007/09/11/for_buyers_carbon_labels_tap_into_worry_on_warming/
www.carbon-label.com

Developing an effective communications plan

To achieve the different objectives discussed above, it will be useful for a business to have a communications plan as part of the GHG management programme. Most communications plans will include the following elements:

- **Clear assignment of responsibility** for communications at all levels.
- **How information is obtained** within the organization, and how the quality of that information will be assured.
- **How and where information will be communicated**. Some examples include: the company website, annual report, sustainability report, staff meetings, internal newsletters and intranet, advertising, news releases, in stores, at conferences, on invoices and customer receipts, on labels and on company vehicles.
- **Mechanisms to ensure consistent communication** across the organization.
- **Opportunities to capture valuable feedback from employees, customers and other stakeholders,** and continually improve the programme.
- **Assignment of responsibility for meeting reporting requirements** for government or voluntary GHG programmes, if applicable.

Principles of good communications for environmental initiatives

There are several principles of good communications for environmental initiatives that apply equally well to GHG management programmes:

- **'Inside out'.** Communicate internally first. This lays a good foundation for both the programme and subsequent external communications. Make sure that key members across all departments of the business are well informed about the GHG management programme so that they can provide consistent messaging from the business to all of its external points of contact.[16]
- **Authenticity.** Align actions and communications, or 'do what you say you are doing'. This is important both for motivating employees and gaining the trust of customers and other stakeholders. Be frank about challenges and obstacles.

- **Gain the knowledge and expertise required to communicate** the GHG management programme effectively. Valid action can be undermined with enthusiastic but inaccurate communication.
- **Be specific.** Provide details about GHG emissions and activities to address them.
- **Obtain assurance from independent third parties.** A recent survey showed that 70% of UK and US consumers want to see independent third-party verification of business claims about climate change action,[17] and other external groups like investors, NGOs and the media also look for this.
- **Seeing is believing.** While it can be challenging to present visual examples of greenhouse gas reductions, there can be opportunities. For example, an installation of solar panels on a company building, or a fleet of hybrid vehicles is a tangible representation of the GHG management programme.

Talking about sustainability

Britain's advertising watchdog found **Shell International Ltd.** violated industry rules when it implied in a newspaper ad that Shell's operations in Canada's tar sands were 'sustainable'. The Advertising Standards Authority (ASA) noted that readers of the ad would assume the term meant environmentally sustainable, even though Shell claimed it had used the term in a broader sense. The ASA further noted it had seen no evidence that Shell was effectively managing its greenhouse gas emissions from its oil sands projects in order to limit climate change, and that the ad had breached rules relating to substantiation, truthfulness and environmental claims.

SOURCE
www.asa.org.uk/asa/adjudications/Public/ TF_ADJ_44828.htm

What is greenwashing?

Greenwashing occurs when a business or other organization tries to camouflage generally poor performance on environmental issues with some very public, but minor or even false displays of environmental action. Some examples include: making vague or unsubstantiated claims; implying certification or endorsement when none exists; promoting one environmental attribute of a product while neglecting other harmful attributes; claiming the status quo or industry standard as an achievement; reporting aspirations as actions; omitting material information; using scientific data very selectively; and inconsistent private and public positions, like lobbying discreetly against regulation while speaking publicly about the need to take strong, decisive action.

Why should businesses be concerned about greenwashing?

Greenwashing can be very damaging to the reputation of a business, with long-term negative implications for brand value. Even if there is no massive public boycott, a business brand can become devalued over time as consumers, inherently sceptical about businesses claiming to support environmental issues, become informed about dubious claims. As well, in many

countries there are regulatory bodies that act as public watchdogs with respect to advertising claims, and they may have the power to sanction businesses that make false or misleading claims.

How to avoid greenwashing

In sum, communicate with integrity. Consult with stakeholders to find out what they are looking for; be transparent; review all environmental claims carefully; ensure that staff are trained to understand the issues; develop a communications policy that addresses greenwashing; seek independent advice and certification; audit to a standard wherever possible; consider communicating through a reputable third-party organization that can verify and critique your green claims; and go beyond what is legally required.

SOURCES
www.greencapital.org.au/index.php?option=com_content&task=view& id=32&Itemid=131
www.terrachoice.com/Home/Six%20Sins%20of%20Greenwashing

GETTING STARTED

MEASURING GHGs

REDUCING GHGs

OFFSETTING GHGs

COMMUNICATING

MOVING FORWARD

CASE STUDY

Tesco
ENGAGING CUSTOMERS IN CLIMATE SOLUTIONS

Tesco plc is the third largest grocery retailer in the world, with 4,331 stores. It employs 470,000 people in 14 countries, including more than 280,000 employees in the UK alone.

In 2007 Tesco announced that it would undertake a broad initiative to address the climate impact of its business. In addition to reducing the footprint of its operations and supply chain, it aimed to empower its many customers — over 30 million people worldwide shop at Tesco every week — to make climate-friendly choices when filling their shopping baskets.

To demonstrate its own commitment, Tesco began to reduce its transportation emissions and make its stores and distribution centres more energy efficient. Its energy use per square foot in its buildings is now half what it was in 2000.

Tesco also listened to its customers and heard that while they were interested in choosing products that were better for the climate, they lacked the information needed to compare the climate impacts of different products. To help them make informed choices, and to identify opportunities for making further reductions in its supply chain, Tesco engaged with its suppliers to measure carbon footprints for several of its product categories over their entire life cycle. The initial product categories included Tesco brands of orange juice, light bulbs, washing detergents and potatoes.

For the laundry detergent category, it turned out that concentrated liquid detergent had a smaller footprint (600g CO_2e per wash) than either powder (750g) or tablets (850g). The product life-cycle analyses also determined that as much as 80% of a detergent's carbon footprint results from customers using it — for example, from using the detergent in hot water instead of cold.

To translate these findings into useful information for customers, Tesco put labels on its laundry detergents identifying the total carbon footprint associated with production, use and disposal of each package. Tesco also provided customers with carbon comparisons of the different laundry detergent products it had measured. For example, on the package for the tablets it was noted that powder detergent had a lower carbon footprint. And finally, both packaging and point-of-sale materials noted that customers could lower the footprint during use by washing at a lower temperature.

Initial feedback from customers was positive, and Tesco now has carbon labels on more than 100 of its products, with plans to increase this to 500 products. Tesco's example demonstrates how companies can both lead by example and use their ready access to customers to educate and influence day-to-day behaviour, and promote climate solutions.

FOR MORE INFORMATION
www.tesco.com

Tailoring messages to different audiences

Below is a table of possible target audiences for the communications plan for the GHG management programme, including reasons for communicating to these audiences, and the information they should receive.

TABLE 5: TAILORING MESSAGES TO DIFFERENT AUDIENCES

Who is the audience?	Why talk to them?	What should you tell them?
CUSTOMERS AND CLIENTS	Customers want businesses to take global warming more seriously, and they may be more willing to do business with companies that are working on global warming solutions. Businesses can enhance their brand by letting customers know about their actions to reduce their impact, and positioning themselves as innovative leaders.[18]	Customers are looking for more information at the point of sale (e.g. on packaging, displays and invoices) about the climate change impact of products and services they purchase, and what the company is doing to mitigate them.[19] Communications can also highlight actions that can be taken by customers.
EMPLOYEES	Employee participation and buy-in can help make the GHG management programme a success. Their feedback can also be a valuable source of ideas for improvements and innovation. Employees care about the social responsibility of employers, and, in a competitive labour market, letting them know how the company manages its climate impact can be advantageous in attracting new employees to the company or retaining existing employees.	Employees need to understand their contribution to making the GHG management programme a success. Communications should make the broader goals of the programme relevant to their activities. In a larger organization the message should be tailored to different departments. Opportunities for feedback are important.
INVESTORS	Investors are recognizing that managing GHG emissions and climate change-related risk will be a significant success indicator. Some investors are beginning to require corporate disclosure around climate change measures.	Investors are interested in how businesses are managing risk related to climate change, how they are capitalizing on new opportunities, and whether they are positioned to adapt to a new regulatory environment. Investors will also likely be interested in how a business compares to industry averages.
MEDIA	Media can be an important source of information about a business for customers and investors.	Businesses can strengthen their media communications by obtaining third-party assurance of claims, and by being as transparent as possible. It may also be useful to demonstrate that there is a strong business case for the actions that the organization is taking, and that customers and employees are engaged.

TABLE 5: TAILORING MESSAGES TO DIFFERENT AUDIENCES *continued*

Who is the audience?	Why talk to them?	What should you tell them?
SUPPLIERS	Many businesses are working with their supply chains to boost the overall success and scope of their own GHG management programmes, and in the process are strengthening relationships with their suppliers.	Suppliers will need to understand the goals of the GHG management programme in a way that is relevant to them, and how to quantify the emissions associated with the products and services they supply. They may also be receptive to information about how to reduce these emissions, and/or develop low-carbon alternatives.
OTHER BUSINESSES AND ORGANIZATIONS	Businesses may want to communicate about their GHG management programmes to other businesses to show leadership, share insights, and encourage others to take action. Businesses in the same industry might want to encourage others to follow suit so that there is a level playing field.	Opportunities include participating in industry-level initiatives, professional networks, or training sessions. Businesses receiving the information will want to know what was done, what worked, and how much it cost. They will be interested in lessons learned, and will possibly be looking for support and mentoring for their own initiatives.
ENVIRONMENTAL GROUPS AND OTHER STAKEHOLDERS	Businesses that reach out to environmental groups, community associations and other stakeholders create opportunities to receive constructive feedback about their GHG management programmes. Some of these groups might be able to direct businesses towards resources, let them know about related initiatives, and even provide public validation of the programme.	Environmental groups will most likely be interested in the overall environmental performance of the business. These groups will be expecting strong GHG reduction efforts and solid evidence of progress being made. Openly discussing challenges can be one way to solicit helpful feedback.
REGULATORY OR VOLUNTARY REPORTING	If a company's GHG management programme is designed to comply with regulations, voluntary standards or GHG programmes (e.g. WWF Climate Savers), there will likely be reporting requirements.	The reporting requirements will be determined by the regulations or programme, and will likely require third-party assurance. These requirements should be considered early in the design of the GHG management programme so that the necessary information is collected and handled appropriately.

Part 5 has described general principles and strategies for communications in support of greenhouse gas management programmes. Of course, in order to tell a truly compelling story, a business needs a strong greenhouse gas management programme, as discussed in other parts of this guide. However, planning communications carefully can contribute to the success and momentum of the programme, by educating and motivating employees, capturing important feedback from all interested stakeholders, and providing an accurate, accessible and inspiring account of the programme.

 Helpful resources
for communicating effectively

Inside Out: Sustainability Communication Begins in the Workplace
green@work, Summer 2005
www.greenatworkmag.com/gwsubaccess/05summer/fiksel.html

Corporate Responsibility and Sustainable Communications: Who's Listening? Who's Leading? What Matters Most?
by Boston Center for Corporate Citizenship, World Business Council for Sustainable Development, Net Impact, and Edelman
www.edelman.com/expertise/practices/csr/documents/
EdelmanCSR020508Final_000.pdf

Eco-Promising: Communicating the Environmental Credentials of Your Products and Service
by Business for Social Responsibility and Forum for the Future
www.bsr.org/reports/BSR_Eco-promising_April_2008.pdf

What Assures Consumers on Climate Change: Switching on Citizen Power
by AccountAbility and Consumers International
www.consumersinternational.org/Templates/Internal.asp?NodeID=96674

Reputation or Reality? A Discussion Paper on Greenwash & Corporate Sustainability *by Total Environment Centre*
www.greencapital.org.au/index.php?option=com_content&task=view&id=32&Itemid=131

The Six Sins of Greenwashing
by TerraChoice Environmental Marketing Inc.
www.terrachoice.com/Home/Six%20Sins%20of%20Greenwashing

Understanding and Preventing Greenwash: A Business Guide
by BSR and Futerra
www.bsr.org/reports/Understanding_Preventing_Greenwash.pdf

Walking the talk
The Footprint Chronicles™ is an interactive mini-website that tracks the impact of seventeen **Patagonia** products from the design stage to delivery at the company's distribution center in Nevada. Footprint data is also provided for more than 150 products, including energy consumption, distance traveled, waste generated, and CO_2 emissions. The goal is not only to inform and engage customers, but also to influence industry practices. 'We've been in business long enough to know that when we can reduce or eliminate a harm, other businesses will be eager to follow suit.'

SOURCE
www.patagonia.com/web/us/footprint

GETTING STARTED

MEASURING GHGs

REDUCING GHGs

OFFSETTING GHGs

COMMUNICATING

MOVING FORWARD

GETTING STARTED

MEASURING GHGs

REDUCING GHGs

OFFSETTING GHGs

COMMUNICATING

MOVING FORWARD

PART 6

Moving Forward and Overcoming Challenges

Part 6 looks at some common challenges faced by businesses when managing their GHG emissions. It offers possible solutions, and ideas about how to strengthen and expand the GHG management programme.

It is helpful to keep in mind that nearly all businesses will encounter some challenges in managing their GHG emissions, especially at the beginning. Some of the more common challenges – and possible solutions – are discussed below:

• **Time and money.** Implementing an effective GHG management programme requires time and financial resources. However, it's worth noting that a GHG management programme is an investment that can have many benefits for the company down the road. Some of these may include: cost savings associated with energy efficiency, brand enhancement, employee satisfaction, and preparedness for a carbon-constrained economy. Businesses may also be eligible for financial incentive programmes, like those referred to in *Helpful resources for reducing GHG emissions* (page 39). To help avoid time crunches, businesses can plan to implement the programme during periods when workloads are less demanding.

• **Too many options.** Because there are many options when it comes to managing GHG emissions – for example, determining which GHG emissions to measure, where to reduce, and whether to use carbon offsets for a carbon neutral initiative – deciding where to start may seem difficult. In many ways, this 'challenge' is actually an opportunity, as the variety of options available gives businesses a lot of flexibility when designing their GHG management programmes. To winnow down the choices, it may be

Identifying challenges

'The barriers will be different in every company,' says Peter Chantraine of **DuPont Canada**. DuPont is one of the world's largest chemical companies. 'I think the key for energy managers is to figure out what those barriers are and who in the company needs to be engaged in finding a solution. You need to talk to people in your legal department, in finance and taxation, and in operations and plant maintenance.'

SOURCE
www.oee.nrcan.gc.ca/publications/
infosource/pub/cipec/DupontEng.
cfm?attr=24

helpful to first spend some time on the steps outlined in Part 1, *Getting Started: Planning for Success* (page 1), i.e. developing a business case for action, and establishing clear goals based on the business case. The goals and business case can then be used to guide subsequent decisions. Second, a rough cost-benefit analysis of the most attractive options can be performed, and if desired, GHG consultants can assist in developing and refining a list of options. Finally, starting with a simple initiative can be a way to gain experience, help identify further activities, and generate momentum for the programme.

- **A steep learning curve for the emissions inventory.** Many businesses find their first emissions inventory more work than expected. Retrieving and tracking all the information associated with calculating emissions for a business is often time-consuming and even frustrating the first time around. The good news is that the second year will likely be much easier, particularly if the business sets up some systems to ensure that the needed data is captured and recorded in a more accessible way.

- **Lack of internal expertise.** Most businesses do not have any previous experience managing greenhouse gas emissions, and may find it somewhat challenging to get started. There are many resources targeted at businesses, including backgrounders, calculation tools, tips on energy savings and other emission reduction strategies, as well as guidance on carbon offsets. See the *Helpful resources* at the end of each part of this guide, as well as the *Additional resources* (page 71). Many businesses are able to get their GHG management programmes started using resources like these.

However, businesses may also find it useful to engage an accredited GHG consultant or an energy auditor to help them with their programme. Such consultations may also be designed to include internal capacity building for the business, putting into place systems and training employees, so that in the future the business can manage its own GHG emissions with less reliance on outside expertise. There are also growing opportunities for education in GHG measurement and management, ranging from online courses to customized seminars.

Businesses may also find it useful to get involved with networks or associations of businesses and other organizations that are tackling GHG

Engaging employees

The **David Suzuki at Work** programme was designed to help employees bring sustainability into their workplace. It contains engaging activities to motivate staff, promote team-building and increase communications across departments. These activities can tie in with a company's overall GHG management plan, and include reducing energy use in the office, creating a smart transportation plan for staff, and developing a green procurement policy.

SOURCE
www.davidsuzuki.org/NatureChallenge/
at_Work/sign_up.aspx

management and climate change issues. Some businesses have engaged with non-governmental and academic organizations to draw on their specialized expertise and insight. See the *Helpful resources* in Part 1.

· **Motivating employees.** Because employees are key drivers of the GHG management programme, it will be difficult for the programme to succeed if employees are uninterested or, more likely, find themselves too busy to contribute their expertise or time to the programme. For this reason, it is important to identify the potential barriers related to employee participation in the programme, especially where behaviour change is required.

Employees can be educated about sources of GHG emissions within the business and why reductions are important, not only for the environment but also for the longer-term viability and competitiveness of the business. They can be made aware of how individual actions to conserve energy or other resources can add up to be important at the organizational level. For example, a business can inform employees about the cumulative impact of turning off computers or a vehicle anti-idling policy, and then look for creative ways to encourage employee participation.

Employees at all levels can contribute useful and often inexpensive ideas about how to reduce GHG emissions. To capture these ideas it's important to establish channels for regular feedback to ensure that employees charged with making decisions hear from employees on the frontlines. This might be as simple as a suggestion box, opportunities for feedback at team meetings, lunch time seminars, or an online forum on the business intranet, where employees can share experiences and insights.

Businesses should also work to create a broad base of engaged employees across the organization, representing different departments or business units. This will help strengthen and harmonize the implementation of the GHG management programme, and maximize its impact, instead of leaving it up to a sustainability team that then must play the role of enforcer.

One way to encourage ongoing employee participation is to incorporate the tasks associated with GHG management into job descriptions, and to allocate adequate resources in terms of time and

Carbon reduction contest engages employees

Business Objects, an SAP company, is a global leader in providing software solutions to businesses. In 2007, the company ran an internal Carbon Footprint Contest, asking its employees for innovative ideas on how to reduce its climate impact. Over 81 ideas were submitted, and all employees voted to determine which ideas were implemented. The resulting reductions from the ideas implemented will be tallied by the company to determine its net greenhouse gas savings.

Business Objects also engages environmental leaders for information sessions where they share their experiences and suggest ways that employees can make a difference, at work and at home.

SOURCE
www.businessobjects.com

GETTING STARTED | MEASURING GHGs | REDUCING GHGs | OFFSETTING GHGs | COMMUNICATING | MOVING FORWARD

money. Some businesses are now including climate change-related objectives (with key performance indicators) in employee evaluations, and are offering incentive packages and other rewards so that efforts supporting the organization's greenhouse gas management objectives can be recognized.

And finally, employees should be informed about organizational successes related to climate change initiatives, and their role in achieving those successes should be celebrated. This can be done, for example, in internal newsletters, in messages from the CEO, or in any way that fits the culture of a particular business.

· **Integration of GHG management into decision-making processes**. A GHG management programme that is not integrated into daily operations and decision-making processes will often compete with other priorities, and will ultimately be less successful. To avoid this, a strong, well-articulated commitment to the programme at the most senior level is critical. For example, a directive from the CEO that 'Business X is going to reduce its GHG emissions 20% from 2005 levels by 2010' is a clear objective that can be used as the basis for decision-making about GHG management.

Next, responsibility for delivering on the commitment needs to be assigned to senior managers or staff who have the authority and resources to make it a reality, and they will need to communicate with each other and coordinate their action. The expertise and authority required to manage and reduce GHG emissions will often be found in different departments or functions, and coordination is an important aspect of GHG management. Bringing these key people to the table for decision-making will produce a well-integrated programme and will help avoid conflicts with existing production or operational plans.

All decision-makers in the organization will need to understand the GHG implications of any plans that are under consideration. For example, if the company's vehicle fleet is being upgraded, GHG emissions produced by the new vehicles and any related costs or savings should be factored into decision-making.

· **Ongoing monitoring and adjustment**. As with any strategic initiative, ongoing monitoring and adjustment are necessary to ensure the GHG management program can adapt and change with current situations. Monitoring can include the company's GHG emissions, the status of

Cultivating an integrated approach to GHG management

Dole, one of the world's largest suppliers of fruits and vegetables, is working to create a carbon neutral supply chain for its bananas and pineapples from Costa Rica. The team managing the initiative has representation from the research department, the environmental department, the department of logistics, the supply chain, corporate responsibility, and marketing. Dole also plans to work closely with businesses in its supply chain, like local transportation companies in Costa Rica.

SOURCE
www.greenbiz.com/news/2007/10/23/
doles-quest-carbon-neutral-supply-chain

reduction efforts, costs and savings related to the programme, carbon offset pricing and availability, media stories about the programme, employee and customer feedback, and others. Companies can learn from their successes as well as failures, and use them to refine and improve the programme over time. Adjusting to changes within the company will also be important. Most businesses experience staff turnover, and new staff need to be educated about the programme. New leaders in the company should reinforce the executive commitment to the programme to create continuing momentum. As well, companies should be prepared to monitor external factors that can impact their programme, including new technologies, actions by competitors, customer demands and government regulations, as all of these might shape future directions for the GHG management programme.

Helpful resources for moving forward

ENGAGING EMPLOYEES

Top 10 Ways to Motivate Employees on Climate Change
by The Climate Group
www.theclimategroup.org/assets/resources/briefing_note_01_internal_
communications.pdf

Saving Money through Energy Efficiency: A Guide to Implementing an Energy Efficiency Awareness Program
by Natural Resources Canada
oee.nrcan.gc.ca/Publications/infosource/Pub/ici/eii/pdf/eii-awareness.pdf

David Suzuki at Work
by David Suzuki Foundation
www.davidsuzuki.org/NatureChallenge/at_Work/sign_up.aspx

GHG MANAGEMENT AND RELATED TRAINING

GHG Management Institute
www.ghginstitute.org

Leading by example
Munich Re, one of the world's largest reinsurers, is a business that has integrated climate change solutions into its operations. Equally important, it has assumed a leadership role internationally on the issue. Munich Re has committed to make its own global operations carbon neutral by 2012, and has developed innovative insurance products for the renewable energy sector that reduce investor risks and make these projects more attractive. It is also leading efforts to develop a massive solar energy project in northern Africa, and is collaborating with the London School of Economics to advance research on the economic consequences of climate change. Each year Munich Re releases industry reports that highlight the link between human-caused climate change and an increasing risk of severe weather-related events around the world, and it has used these findings to call on governments to adopt effective rules to limit greenhouse gas emissions.

SOURCE
www.munichre.com

Conclusion

Climate change presents a major challenge for the business community. It will profoundly alter the way business is conducted in the future, illustrating how the environment and the economy are inextricably linked. The physical impacts of climate change have the potential to negatively affect economic interests, but, on the other hand, efforts to address climate change can spur innovation, expand markets for clean technologies and create new green collar jobs.

Businesses that don't take action to manage their greenhouse gas emissions, and mitigate their climate-related risks, face an uncertain future. Customers are already beginning to move away from companies that aren't taking climate change seriously, and investors are asking questions about how businesses are adapting their organizations to new demands related to GHG management. Government regulators in many countries are also exploring ways to set a price on carbon, and some jurisdictions have already enacted legislation to limit emissions. All of these challenges – combined with high energy prices that affect the bottom line – are providing clear economic incentives for businesses to reduce their GHG emissions.

For businesses that act now, climate change presents enormous opportunities. Examples from around the world show that businesses that manage their greenhouse gas emissions and develop climate-friendly products and services are not only saving substantial amounts of money, but are also realizing other competitive advantages. These include new customers and markets, loyal and motivated employees, strengthened relationships with suppliers and increased operational efficiencies.

Although managing greenhouse gas emissions is still new territory for many businesses, it is expected to become a mainstream practice in the near future. Businesses that show leadership by acting now to measure, reduce and offset their GHG emissions will be able to limit their exposure to the risks, and at the same time take advantage of important opportunities. From a broader perspective, these businesses will also be lending their workforce and their creativity to help solve the problem of climate change.

Additional resources

The resources below supplement the resources at the end of each part of this guide. For a current, comprehensive list of climate change resources for businesses see: www.davidsuzuki.org/Climate_Change/business.asp

BOOKS ABOUT BUSINESS AND SUSTAINABILITY

Ricardo Bayon, Amanda Hawn, Katherine Hamilton (2007) *Voluntary Carbon Markets: A Business Guide to What They Are and How They Work*, Earthscan Publications Ltd.

Daniel C. Esty and Andrew S. Winston (2006) *Green to Gold: How Smart Companies Use Environmental Strategy to Innovate, Create Value and Build Competitive Advantage*, Yale University Press.

Paul Hawken (1994) *The Ecology of Commerce*, Collins.

Andrew Hoffman (2007) *Carbon Strategies: How Leading Companies Are Reducing Their Climate Change Footprint*, University of Michigan Press.

Bob Willard (2002) *The Sustainability Advantage – Seven Business Case Benefits of a Triple Bottom Line*, New Society Publishers.

BACKGROUND ON BUSINESS AND CLIMATE CHANGE

The Climate Group
www.theclimategroup.org

ClimateBiz – The Business Resource for Climate Management
www.climatebiz.com

Green Business Resources *by Natural Resources Defense Council*
www.nrdc.org/enterprise

Climate Changes Your Business
by KPMG
www.kpmg.nl/site.asp?id=40378&process_mode=mode_doc&doc_id=45618

Getting Ahead of the Curve: Corporate Strategies that Address Climate Change
by the Pew Center on Global Climate Change
www.pewclimate.org/global-warming-in-depth/all_reports/corporate_strategies

CEO Briefing: Carbon Crunch, Meeting the Cost
by the United Nations Environment Programme
www.unepfi.org/fileadmin/documents/CEObriefing_carbon_crunch.pdf

Beyond Neutrality: Moving Your Company Toward Climate Leadership
by Business for Social Responsibility
www.bsr.org/reports/BSR_Beyond-Neutrality.pdf

The Carbon Trust
www.carbontrust.co.uk

Ahead of the Curve: Business Responds to Climate Change
by Sea Studios Foundation
http://seastudios.org/ahead.php

BUSINESS RISKS RELATED TO CLIMATE CHANGE

Adapting to Climate Change: Business Planning, Risk Management
and Emergency Preparedness
by The Conference Board of Canada
www.conferenceboard.ca/documents.asp?rnext=2452

Adapting to Climate Change: A Business Approach
by the Pew Center on Global Climate Change
www.pewclimate.org/business-adaptation

Investor Network on Climate Risk *by Ceres*
www.incr.com

Global Climate Change Impacts in the United States
by United States Global Change Research Program
www.globalchange.gov/publications/reports/scientific-assessments/us-impacts

CLIMATE CHANGE SCIENCE AND POLICY

Intergovernmental Panel on Climate Change
www.ipcc.ch

United Nations Framework Convention on Climate Change
www.unfccc.int

Stern Review: The Economics of Climate Change
www.occ.gov.uk/activities/stern.htm

Water, Energy and Climate Change: A Contribution from the Business
Community *by World Business Council for Sustainable Development*
www.unwater.org/downloads/WaterEnergyandClimateChange.pdf

CARBON MARKET SURVEYS AND ANALYSIS

State of the Voluntary Carbon Markets
by Ecosystem Marketplace
www.ecosystemmarketplace.com

State and Trends of the Carbon Market
by The World Bank
http://carbonfinance.org

Glossary

Absolute target: A target defined by a reduction in absolute (total) emissions over time, e.g. the reduction of CO_2 emissions by 20% below 2000 levels by 2010. Compare *intensity target*.

Activity data: Data from activities that generate emissions, such as driving a vehicle or using electricity, in units that allow for emissions to be calculated (e.g. kilometres driven, litres of fuel used, kilowatt hours, etc.).

Additionality: Refers to an essential characteristic of carbon offsets, i.e. that they must result from emission reduction activities carried out because of the incentives associated with the existence of the carbon market, and not be the result of 'business as usual' activities. A variety of tests have been developed to assess the additionality of offset projects.[20]

Base year: A specific year (or an average over multiple years) against which an organization's emissions can be tracked over time.

Carbon calculator: *See GHG emission calculation tool.*

Carbon dioxide (CO_2): A naturally occurring gas, and also a by-product of burning fossil fuels and biomass, land-use changes and industrial processes. It is the greenhouse gas responsible for most of the Earth's warming caused by human activity. See also *carbon dioxide equivalent*.

Carbon dioxide equivalent (CO_2e): The universal unit of measurement used to indicate the global warming potential of each of the six greenhouse gases so that their relative climate impact can be compared and overall climate impact aggregated. The CO_2e quantity of any greenhouse gas is the amount of carbon dioxide that would produce the equivalent global warming potential. See also *global warming potential*.

Carbon footprint: The greenhouse gas emissions associated with a particular individual, organization, company, other entity or activity. These may include direct emissions such as those from driving a car or burning fuel to heat a building and/or indirect emissions such as those from flying in a commercial airplane or using electricity purchased from a utility.

Carbon neutral: Used to signify that an organization or individual has reduced the net climate impact of their operations or activities to zero, usually after purchasing offsets in a quantity equal to their total emissions after reduction efforts. For example, a business with total emissions of 100 tonnes (after its own direct reductions) would purchase 100 tonnes of offsets to become carbon neutral.

Carbon neutral product or service: A product or service for which all the significant greenhouse gas emissions associated with bringing that product or service to market are offset. See also *life-cycle analysis*.

Carbon offset: A reduction in greenhouse gas emissions created by one party that can be purchased and used to compensate for (offset) the greenhouse gas emissions of another party. Carbon offsets are quantified in metric tonnes of CO_2e reductions. They may be purchased on a voluntary basis or to meet regulatory requirements. The effectiveness of carbon offsets in creating real reductions in greenhouse gas emissions depends on whether they meet important quality criteria.

Carbon offset standard: A standard that helps to ensure that carbon offset projects meet certain quality criteria, such as additionality and third-party auditing. A number of offset standards exist for the voluntary and compliance markets, and each has a slightly different focus and set of requirements. See also *Gold Standard, The.*

Climate change: A change of climate attributed directly or indirectly to human activity that alters the composition of the global atmosphere and which is in addition to natural climate variability observed over comparable time periods. See also *global warming.*

Climate leadership team: The people within an organization tasked with developing and implementing a GHG management programme.

Climate neutral: See *carbon neutral.*

Control: For the purposes of the GHG Protocol, a company's ability to direct the policies of another operation, even when it does not own the operation. More specifically, it is defined as either operational control (where the organization or one of its subsidiaries has the full authority to introduce and implement its operating policies at the operation) or financial control (where the organization has the ability to direct the financial and operating policies of the operation with a view to gaining economic benefits from its activities). The concept of control was developed to assign responsibility for GHG emissions.[21]

Corporate social responsibility (CSR): The balanced integration of social, economic and environmental considerations into business decision-making, and the engagement of stakeholders in that process.

Direct emissions: According to the GHG Protocol, these are GHG emissions from sources that are owned or controlled by an organization. See also *scope 1 emissions.*

Double-counting: This occurs when two or more entities claim ownership of the same emissions or reductions.

Downstream reductions: Reductions in the GHG emissions that occur after a product is manufactured and sold to customers, or after a service has been performed for a customer. Examples include the GHG emissions associated with delivery by another organization, customer travel to purchase products or obtain services, the use of products and services, and final disposal of products.

Emission factor: A coefficient (e.g. grams of carbon dioxide emitted per litre of fuel consumed) used to calculate GHG emissions.

Emission reductions: The measurable reduction of greenhouse gas emissions from a specified activity or over a specified area and a specified period of time.

Emissions inventory: A list of an organization's GHG sources with emissions quantified.

Emissions trading: A regulated system that sets an overall emissions limit (or cap), allocates emissions allowances to participants, and allows them to trade these allowances, and sometimes carbon offsets, with each other to meet their individual targets. Also called a 'cap and trade system'.

Energy conservation: Reduction or elimination of unnecessary energy use.

Energy efficiency: The rate at which a machine or other equipment uses energy to perform its function.

Fossil fuels: Fuels like petroleum, coal or gas which originate in the earth as hydrocarbon deposits, and generate greenhouse gas emissions when burned.

Fugitive emissions: Planned or unplanned emissions that result from leakage during the production, processing, distribution, storage and use of fuels and other chemicals.

GHG: See *greenhouse gas*.

GHG emission calculation tool: A tool (often found on websites) that allows users to calculate how much carbon dioxide or other greenhouse gases are emitted from various activities, such as air travel.

GHG emissions: The release of greenhouse gases into the atmosphere, either intentionally or unintentionally.

GHG management programme: A programme undertaken by a business or other organization that includes such activities as measuring, reducing, offsetting and reporting its greenhouse gas emissions.

GHG programme: Any voluntary or mandatory international, national, sub-national, regional, government, or non-governmental programme that registers, certifies, regulates or manages greenhouse gas emissions from organizations.[22]

GHG source: Any physical unit or process which releases greenhouse gases into the atmosphere; for example, electricity generation.

Global warming: The gradual increase, observed or projected, in global surface temperature as one of the consequences of an accumulation of greenhouse gases in the atmosphere. See also *climate change*.

Global warming potential (GWP): A measure of how much a given amount of greenhouse gas is estimated to contribute to global warming, relative to the same amount (by weight) of carbon dioxide (whose GWP is by definition 1). See also *carbon dioxide equivalent (CO_2e)*.

Gold Standard, The: A rigorous standard for carbon offsets that requires that offsets be generated by renewable energy or energy efficiency projects that also promote sustainable development in the host communities. Gold Standard offset projects are tested for environmental quality by UN-accredited auditors according to recognized methodologies.

Green power: See *renewable energy*.

Greenhouse gas (GHG): Any natural or man-made gas that absorbs infrared radiation in the atmosphere. The six greenhouse gases that are covered by the Kyoto Protocol are: carbon dioxide (CO_2); methane (CH_4); nitrous oxide (N_2O); hydrofluorocarbons (HFCs); perfluorocarbons (PFCs); and sulphur hexafluoride (SF_6). Others not covered include water vapour, because it is very short-lived, and ozone. See also *carbon dioxide equivalent (CO_2e)*.

Greenhouse Gas Protocol Initiative: A multi-stakeholder collaboration convened by the World Resources Institute and World Business Council for Sustainable Development to design, develop and promote the use of greenhouse gas accounting and reporting standards for business. It comprises two related standards: the GHG Protocol Corporate Accounting and Reporting Standard, and the GHG Protocol for Project Accounting.[23]

HVAC: Heating, ventilation and air conditioning systems.

Indirect emissions: According to the GHG Protocol, emissions that are a consequence of the operations of an organization, but occur at sources owned or controlled by another entity.[24] See also s*cope 2 emissions* and *scope 3 emissions*.

Intensity target: A target defined by a reduction in the ratio of GHG emissions to a business metric over time: e.g. the reduction of CO_2 per unit of production by 10% between 2000 and 2008.

Intergovernmental Panel on Climate Change (IPCC): International body of climate change scientists set up by the World Meteorological Organization (WMO) and by the United Nations Environment Programme (UNEP). Its role is to assess the scientific, technical and socio-economic information relevant to the understanding of the risk of human-induced climate change.

Inventory: See *emissions inventory*.

ISO 14064: The ISO 14064 Greenhouse Gases is a voluntary series of standards developed with stakeholders from industry, government, NGOs and service professionals. ISO 14064 is designed to help organizations and governments in measuring, reporting and verifying greenhouse gas emissions.[25]

Kyoto Protocol: An international protocol to the United Nations Framework Convention on Climate Change (UNFCCC) that requires industrialized country signatories to meet reduction targets of greenhouse gas emissions relative to their 1990 levels during the period of 2008–2012.

LED: Light-emitting diodes. Due to their high efficiency and long life, LED lights are increasingly replacing incandescent and compact fluorescent lights.

Life-cycle analysis (LCA): An assessment of the sum of a product or service's environmental impacts (such as greenhouse gas emissions) over its entire life cycle, including resource extraction, production, use and disposal.

Metric tonne CO_2e: The usual unit of measurement for greenhouse gas emissions. One metric tonne = 1000 kilograms = 2204.6 pounds = 1.1023 US short tons = 0.9842 UK long tons. See also *carbon dioxide equivalent*.

Offsets: See *carbon offsets*.

Operational boundary: According to the GHG Protocol, the boundary defining which direct and indirect emission sources from company operations within the organizational boundary will be included in the emissions inventory.

Organizational boundary: According to the GHG Protocol, the boundary defining the company operations to be included in the emissions inventory, based on ownership and equity share or control. See also *control*.

Permanence: An aspect of offset quality that refers to the durability of the climate benefit from an offset project, i.e. whether there is any risk of reversal.

Purchased electricity, heat or steam: Electricity, heat or steam used by an organization but generated by another company. See also *scope 2 emissions*.

Renewable energy: Energy from sources that are essentially inexhaustible, such as wind, hydropower, solar, geothermal, biomass, etc., and which also emit fewer GHG emissions than the burning of fossil fuels. Electricity from renewable sources is sometimes called 'green power'.

Renewable Energy Certificate (REC): A REC is a certificate issued when one megawatt-hour of electricity is generated and delivered to the grid from a qualifying renewable energy source, such as wind or solar. RECs are sold by utilities and green energy developers to consumers who want to support renewable energy, even though they may not receive electricity directly from the renewable energy source. See box on page 34.

Scope: Refers to the emissions included in an organization's inventory according to the operational boundary it has drawn. See also *operational boundary.*

Scope 1 emissions: According to the GHG Protocol, the greenhouse gas emissions from sources that are owned or controlled by an organization. It includes, for example, the fuel used by a company to power its vehicles, generate heat and run manufacturing equipment. See Table 2 (page 12).

Scope 2 emissions: According to the GHG Protocol, an organization's indirect greenhouse gas emissions from purchased electricity, heat or steam. See Table 2 (page 12). See also *purchased electricity, heat or steam.*

Scope 3 emissions: According to the GHG Protocol, an organization's indirect greenhouse gas emissions, i.e. those from sources not owned or controlled by the organization, other than those covered in scope 2. Examples include supply chain emissions, transportation in vehicles not owned or controlled by the organization, the use of its products and the disposal of its waste. See Table 2 (page 12).

Supply chain emissions: GHG emissions associated with suppliers. See also *upstream emissions.*

Upstream emissions: Emissions from activities that take place before a product or service reaches an organization. Some examples include the extraction of raw materials, and the production of components.

Verification: An independent assessment by a qualified party of the reliability of a greenhouse gas inventory or the reductions from a carbon offset project.

Voluntary carbon market: The segment of the carbon market that includes all carbon offset transactions that are not part of government-regulated compliance schemes. It serves individuals, businesses and other organizations that voluntarily choose to take responsibility for their climate impact. See box on page 51.

ENDNOTES

1 Although the meanings of these terms differ slightly, they are often used interchangeably. See *Glossary* for definitions.

2 *Climate Change 2007: Synthesis Report. Contribution of Working Groups I, II and III to the Fourth Assessment Report of the Intergovernmental Panel on Climate Change* (2007) (Core Writing Team, R.K Pachauri and A. Reisinger (eds.)), Geneva, Switzerland: IPCC. See also *From Impacts to Adaptation: Canada in a Changing Climate 2007* (2007) Donald Lemmen, Fiona Warren, Elizabeth Bush and Jacinthe Lacroix (eds.) Ottawa, Canada: Natural Resources Canada.

3 Nicholas Stern (2006) *The Economics of Climate Change: The Stern Review,* Cambridge, UK: Cambridge University Press. www.guardian.co.uk/environment/2008/jun/26/climatechange.scienceofclimatechange.

4 James Pomfret, 'Risks of global warming greater than financial crisis: Stern' (Reuters) 27 October 2008. www.reuters.com/article/environmentNews/idUSTRE49Q19120081027

5 World Resources Institute and World Business Council for Sustainable Development (2004) *The Greenhouse Gas Protocol: A Corporate Accounting and Reporting Standard*, revised edition, Washington, DC: World Resources Institute. See also Samantha Putt del Pino, Ryan Levison and John Larsen (2006) *Hot Climate, Cool Commerce: A Service Sector Guide to Greenhouse Gas Management*, Washington, DC: World Resources Institute, developed specifically for service sector companies.

6 World Resources Institute and World Business Council for Sustainable Development (2004) *The Greenhouse Gas Protocol: A Corporate Accounting and Reporting Standard*, revised edition, Washington, DC: World Resources Institute, p. 16.

7 See Samantha Putt del Pino, Ryan Levinson and John Larsen (2006) *Hot Climate Cool Commerce: A Service Sector Guide to Greenhouse Gas Management,* Washington, DC: World Resources Institute, p. 47.

8 Canadian Business for Social Responsibility (2007), *The Climate Change Guide, Corporate Canada: Responsible Business Action on Climate Change*, Toronto: Canadian Business for Social Responsibility, p. 41.

9 An energy auditor that has been engaged to identify opportunities to make reductions in energy efficiency should also be able to provide detailed information about the investment required and payback periods on the investment.

10 For more information about building recommissioning, see the US Energy Star publication *Recommissioning* at www.energystar.gov/ia/business/BUM_recommissioning.pdf

11 Greenhouse gas (GHG) emissions by source, 2004. In UNEP/GRID-Arendal Maps and Graphics Library. http://maps.grida.no/go/graphic/greenhouse-gas-ghg-emissions-by-source-2004.

12 See Aimee McKane and Joseph C. Ghislain (2006) 'Energy Efficiency as Industrial Management Practice: The Ford Production System and Institutionalizing Energy Efficiency.' SAE International, cited in Business for Social Responsibility (2006) *A Three-Pronged Approach to Corporate Climate Strategy*, San Francisco: Business for Social Responsibility, p. 21.

13 Based on calculations from the Environmental Defense paper calculator, www.edf.org/papercalculator

14 www.cserv.gov.bc.ca/ministry/docs/climate_action_charter.pdf

15 It should be noted that the Greenhouse Gas Protocol pertains to emissions inventories and does not address carbon neutral initiatives.

16 For more on this topic, see Joseph Fiksel, Robert A. Axelrod and Susan Russell, 'Inside Out: Sustainability Communication Begins in the Workplace', green@work, Summer 2005. www.greenatworkmag.com/gwsubaccess/05summer/fiksel.html.

17 AccountAbility and Consumers International (2007) *What Assures Consumers on Climate Change: Switching on Citizen Power*, London: AccountAbility and Consumers International, p. 9.

18 A recent study has shown that brand can represent up to 60% of a company's worth. See Adrian Davis and Lucinda Spicer, 'An International Perspective on Brand Valuation and Management', PricewaterhouseCoopers UK (article in *IP Value 2004: Building and Enforcing Intellectual Property Value*, Global White Page).

19 AccountAbility and Consumers International (2007) *What Assures Consumers on Climate Change: Switching on Citizen Power*, London: AccountAbility and Consumers International, p. 24.

20 *A Consumer's Guide to Retail Carbon Offset Providers* (2006) Clean Air–Cool Planet, p. viii.

21 World Resources Institute and World Business Council for Sustainable Development (2004) *The Greenhouse Gas Protocol: A Corporate Accounting and Reporting Standard,* revised edition.

22 Ibid., definition modified for use in this guide.

23 Ibid.

24 Ibid., definition modified for use in this guide.

25 Canadian Standards Association, www.csa.ca.

INDEX